NATURAL ENVIRONMENT RESEARCH COUNCIL

Institute of Terrestrial Ecology

Atlas of the

Lichens

of the British Isles

Volume I

Edited by

M R D   Seaward
and
C J B Hitch

UNIVERSITY OF BRADFORD

1982

Printed in Great Britain by
NERC/SERC Reprographic Services, Swindon
© NERC Copyright 1982
Published in 1982 by
Institute of Terrestrial Ecology
68 Hills Road
Cambridge
CB2 1LA
0223 (Cambridge) 69745

ISBN 0 904282 57 0

The cover shows *Haematomma ventosum* (L.) Massal., (map 56), an upland species of exposed siliceous rocks.

Photograph F Dobson

These maps are published as part of the Biological Records Centre's programme for recording the distribution of the fauna and flora of the British Isles. The Biological Records Centre is operated by the Institute of Terrestrial Ecology (Natural Environment Research Council) and receives financial support from the Nature Conservancy Council.

The Institute of Terrestrial Ecology (ITE) was established in 1973 from the former Nature Conservancy's research stations and staff, joined later by the Institute of Tree Biology and the Culture Centre of Algae and Protozoa. ITE contributes to and draws upon the collective knowledge of the fourteen sister institutes which make up the Natural Environment Research Council, spanning all the environmental sciences.

The Institute studies the factors determining the structure, composition and processes of land and freshwater systems, and of individual plant and animal species. It is developing a sounder scientific basis for predicting and modelling environmental trends arising from natural or man-made change. The results of this research are available to those responsible for the protection, management and wise use of our natural resources.

Nearly half of ITE's work is research commissioned by customers, such as the Nature Conservancy Council who require information for wildlife conservation, the Department of Energy, the Department of the Environment and the EEC. The remainder is fundamental research supported by NERC.

ITE's expertise is widely used by international organisations in overseas projects and programmes of research.

M R D Seaward
Postgraduate School of Studies in Environmental Science
University of Bradford
Bradford, West Yorkshire BD7 1DP
0274 (Bradford) 33466

C J B Hitch
Former Research Fellow
University of Bradford.

DEDICATED TO
PETER JAMES,
DAVID HAWKSWORTH, FRANCIS ROSE,
BRIAN COPPINS, PAULINE TOPHAM,
TONY FLETCHER, OLIVER GILBERT
AND THE MANY OTHER MEMBERS
OF THE BRITISH LICHEN SOCIETY
WITHOUT WHOM THIS MAPPING
PROGRAMME WOULD NOT
HAVE BEEN POSSIBLE

INTRODUCTION

This Atlas aims to show as clearly as possible the present-day distribution of British lichens, to provide an informed commentary on their chorology and ecology, and to provide a base-line against which future change can be assessed. Such a survey became a possibility soon after the formation of the British Lichen Society in 1958. A mapping scheme, first proposed in 1962 by a small working party, became reality in 1963; one of the members of that group, M.R.D. Seaward, became the Society's Mapping Recorder, a position he has held since that date. By 1972 the volume of data amassed justified publication of maps, some of which were incorporated into monographic studies and others were published with detailed rubrics as a series in the *Lichenologist* (see Appendix A).

Major grants for this research programme, which included a two-year post-doctoral research fellowship for one of us (CJBH), were generously awarded by the Natural Environment Research Council. Additional funding for fieldwork was provided by a variety of sources, including the World Wildlife Fund and the Praeger Committee of the Royal Irish Academy, and a grant from IBM was used to produce essential data-transfer sheets (see below).

METHODS

Standard mapping cards (often several for a particular 10 km x 10 km recording unit) listing field records from 1960 onwards have been filed since 1963. These record cards supplement the information shown on our lichen maps, often including date(s), habitat, precise locality, vice-county and name of recorder(s), and occasionally annotations such as status and fertility of species; where habitats of particular interest and/or complexity are involved, additional cards are on file. Such cards provide a wealth of data for phytogeographers, conservationists, compilers of floras and keepers of regional data banks.

Records for 700 taxa were transferred onto specially-designed data-transfer sheets which facilitated easy conversion onto 80-column computer cards, and thence onto magnetic tape and disc. Using programmes devised by the University of Bradford Computer Centre, lists of coded data and readable species lists for each 10 km x 10 km grid square were prepared, and this information was validated visually. Readable location print-outs (4-figure grid references) for each of the 700 taxa listed on the data-transfer sheet were then prepared from the validated coded information. The locational data were employed in the preparation of preliminary distribution maps of each taxon by means of a line-printer; those for Ireland were prepared separately.

250 of these maps were selected for possible inclusion in Volume 1 of the Atlas, and were circulated to twenty British lichenologists. The maps proved most valuable as working documents, and were returned, duly annotated, to include further modern records and as many pre-1960 records as possible. The returned maps, together with all other programmes, are being up-dated continually by MOP computer work.

176 of the 250 maps circulated were eventually selected for publication. Maps in a format suitable for publication were produced by ourselves, with occasional help from university students, and a rubric for each was prepared, mainly by Mr P.W. James, Dr D.L. Hawksworth, Dr F. Rose, Mr B J. Coppins and Dr A. Fletcher.

SOURCES

A detailed analysis of published sources was also undertaken to complement the above work; many of the data assembled have been incorporated into the bibliographical survey by Hawksworth and Seaward (1977), which contains more than 2700 citations cross-referenced to vice-counties. Detailed analysis of major and minor herbaria and of manuscript lists, etc. have also been made in conjunction with the mapping programme, and accounts of some of these analyses have been published. We are indebted to the many members of the British Lichen Society and others who have undertaken such analyses, and to the keepers and curators who have allowed access to their collections, notably the British Museum (BM), Edinburgh (E), Dublin (DBN), Belfast (B) and Cardiff (NMW), and also to the help afforded by private owners of documents and herbarium material. Further details of this work, including a list of herbaria consulted, are to be found in Hawksworth and Seaward (1977).

INTERPRETATION

The 176 maps in this volume are of species known to have been mapped comprehensively. Records of all species continue to flow in: additional volumes are intended, but in the meantime more than 100,000 records covering the remaining British taxa are available on computer files, and similar numbers of additional records are at present housed on data files, in the Computer Centre and the School of Environmental Science, University of Bradford.

Study of the maps reveals important aspects of lichen chorology and ecology. The influence of climate, for example, is well shown by *Parmelia laevigata* (map 100) = Atlantic, *Anaptychia ciliaris* (map 4) = eastern, *Cetraria sepincola* (map 37) = boreal, *Parmelia soredians* (map 109) = southern, etc, (see Coppins, 1976). Other ecological factors important in controlling lichen distribution are maritime influence (e.g. *Haematomma ventosum,* map 56), altitude (e.g. *Cetraria nivalis,* map 36), and pH of substrate (e.g. *Acarospora fuscata* (map 1) and *Solorina saccata* (map 152) being found on acidic and calcareous rocks respectively).

Recent change effected by such dynamic influences as the spread of sulphur dioxide air pollution (see Fig. 2) are highlighted by the inclusion of old records which demonstrate the decline of species such as *Parmelia caperata* (map 92). Conversely, several species, such as *Lecanora conizaeoides* (map 64) and *Parmeliopsis ambigua* (map 116), have spread under the same influence. However, in areas where levels of sulphur dioxide pollution are known to be decreasing, the situation is more complex: it is expected that base-line distributional data in this Atlas can be used to evaluate amelioration.

Decline due to air pollution is sometimes difficult to separate from the effects of habitat destruction, but it is now believed that the latter factor is largely responsible for the disappearance of such species as *Lobaria pulmonaria* (map 75)—an old woodland indicator—from much of England and the central lowland areas of Scotland. More recently the loss of mature elm trees through Dutch Elm Disease has had a dramatic effect, and host-specific species such as *Caloplaca luteoalba* (map 24) face extinction in the British Isles.

Finally, the ecological amplitude of many lichen species cannot be determined from distribution maps alone; for example, *Diploicia canescens* (map 47) is essentially epiphytic in the south but saxicolous in the north. Furthermore, maps show the range of species, but not their frequency; for example, records of *Ramalina farinacea* (map 140) from large areas of central and eastern England are often based only on small isolated thalli, which may be either relict or newly-established.

NOMENCLATURE

The nomenclature of species and communities are according to Hawksworth *et al.* (1980) and James *et al.* (1977) respectively.

SYMBOLS

Two symbols are used on the maps to differentiate the date of records: a black spot for records from 1960 onwards, and an open circle for pre-1960 records.

ACKNOWLEDGEMENTS

The enormous task of mapping over the past seventeen years was made possible only with the help and co-operation of many amateur naturalists and professional biologists. A list of all those who have participated in the fieldwork is provided (Appendix B)—to all of them we owe our heartfelt thanks. We are also grateful to the many individuals and institutions (see above), far too numerous to list, who have given, and continue to give, their support to this programme. We thank J. Heath and C.D. Preston of the Institute of Terrestrial Ecology's Biological Record Centre for their help in the production of this volume.

FUTURE RECORDING

The British Lichen Society mapping scheme continues, not only to prepare further volumes of the Atlas but also to up-date those maps already published. New recorders are always welcome. Further details regarding the scheme may be obtained from the Mapping Recorder, Dr M.R.D. Seaward, School of Environmental Science, University of Bradford, Bradford BD7 1DP.

REFERENCES

Coppins, B.J. 1976. Distribution patterns shown by epiphytic lichens in the British Isles. In: *Lichenology, progress and problems,* edited by D.H. Brown, D.L. Hawksworth and R.H. Bailey, 249-278. London: Academic Press.

Hawksworth, D.L., James, P.W. & Coppins, B.J. 1980. Checklist of British lichen-forming, lichenicolous and allied fungi. *Lichenologist,* 12, 1-115.

Hawksworth, D.L. & Seaward, M.R.D. 1977. *Lichenology in the British Isles 1568-1975. An historical and bibliographical survey.* Richmond: Richmond Publishing.

James, P.W., Hawksworth, D.L. & Rose, F. 1977. Lichen communities in the British Isles: a preliminary conspectus. In: *Lichen ecology,* edited by M.R.D. Seaward, 295-413. London: Academic Press.

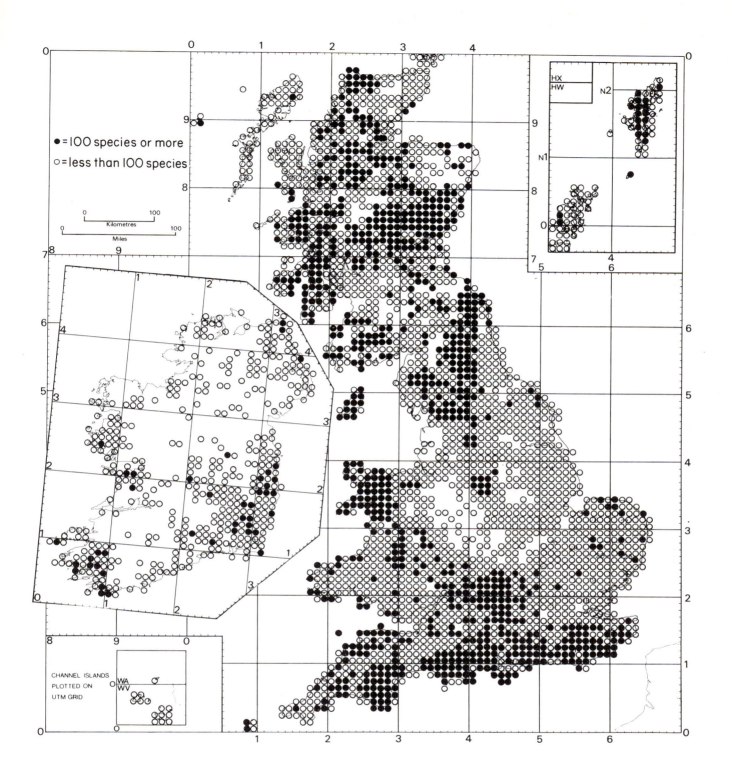

Fig. 1 Completed cards based on fieldwork undertaken since 1960 have been received for 2986 ten-kilometre grid squares. Of these, 50 list five or less records and have been omitted from the above map and the following analysis: 902 cards (including 43 from Ireland) have over 100 records; 858 cards (including 112 from Ireland) have 50-100 records; and 1176 cards (including 257 from Ireland) have less than 50 records. Over 78% of the British Isles has received some form of coverage: England, Wales and Scotland, with few exceptions (e.g. Lancashire and the north-west Midlands), have received good coverage, but Ireland, with the exceptions of reasonably large areas of the south-west and south-east, is still disappointingly underworked. These poorly-worked areas should be taken into account when interpreting maps of lichens, especially of those species with a more widespread distribution. Intermediate maps illustrating the progress of the British Lichen Society's mapping scheme for the whole of the British Isles are to be found in the *Lichenologist* **5**, 464-466 (1973), **7**, 180 (1975) and **11**, 324 (1979), and for Ireland in the *Ir.Nat.J.,* **18**, 335-336 (1976) and **20**, 164-165 (1980).

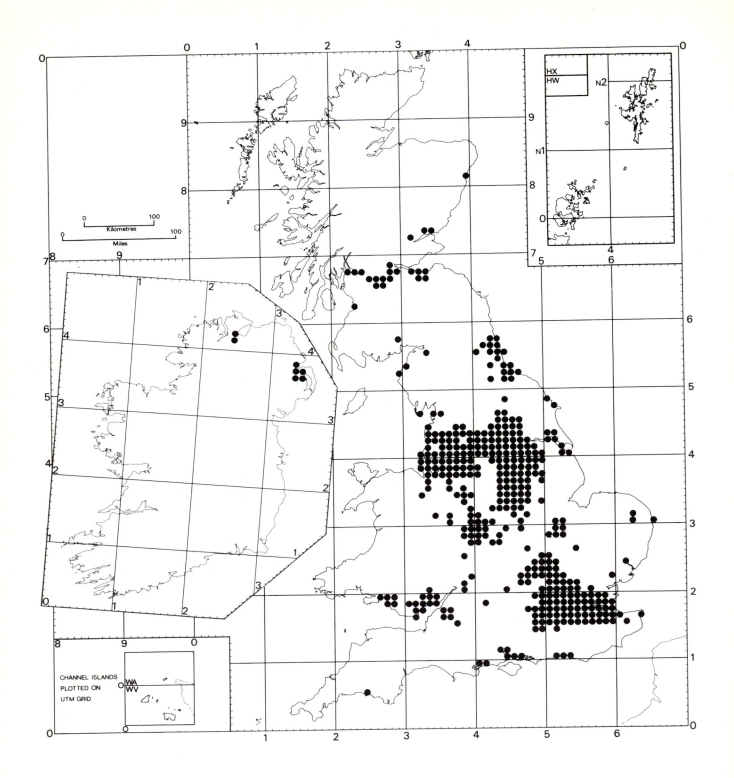

Fig. 2 Map illustrating mean winter sulphur dioxide levels of more than *c.* 100 μg m⁻³ in the United Kingdom, based on a 1972-73 contour map prepared by Warren Spring Laboratory, Stevenage; comparable data for the Irish Republic are not available. The measurements are derived from a national monitoring programme, but the majority of the gauges are sited in urban and industrial areas: probably no more than 5% of the instrumentation (*c.* 50 gauges) is located in rural areas, which occupy about 95% of the land surface. Lichen distributional data can therefore be used to complement these monitoring programmes in rural areas (see Hawksworth, D.L. and Rose, F., 1976, *Lichens as pollution monitors,* London: Edward Arnold). Since sulphur dioxide concentration is known to be a major factor shaping lichen distribution, the above pollution information should assist in the interpretation of many of the following maps. Urban sulphur dioxide levels have fallen significantly since the original contour map was prepared, but changes in rural areas have been less pronounced. However, lichens are subject to time-lag in their response to a decrease in pollution, and the above map is still highly relevant. The role of other environmental factors influencing lichen distribution can be determined by the use of specially prepared overlays (*Overlays of environmental and other factors for use with Biological Records Centre distribution maps,* 1978, Cambridge: Institute of Terrestrial Ecology).

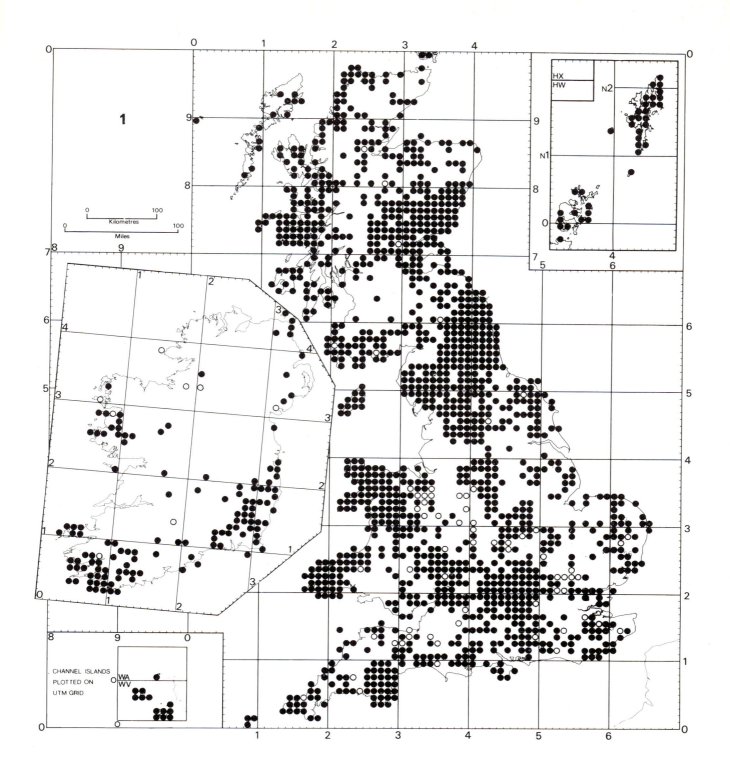

1   *Acarospora fuscata* (Nyl.) Arnold

A species of coarse-grained, nutrient-enriched, sunny, siliceous rocks and walls, in coastal and inland areas. Frequent on man-made substrates in lowland Britain. In upland areas, a characteristic species of birds' perching stones. Often under-recorded in the past, but easily distinguished from other British species of the genus by the C+ red reaction of the thallus. Moderately tolerant of air pollution, but absent from areas with mean sulphur dioxide levels exceeding 65 μg m^{-3} and declining in areas of intensive agriculture due to sensitivity to inorganic fertilizer sprays. An important member of the *Parmelion conspersae.* Widespread in Europe, but distribution elsewhere uncertain.

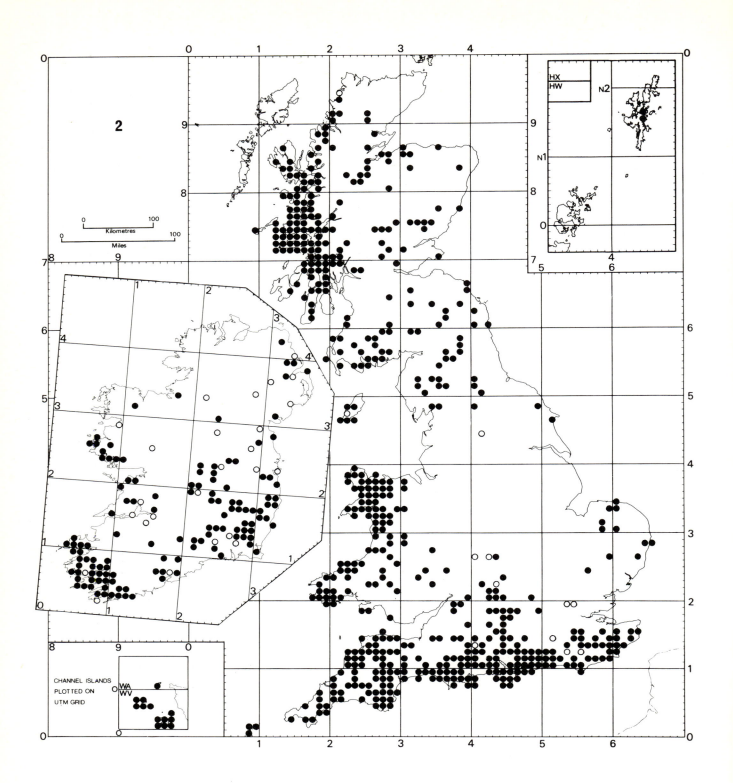

2 *Acrocordia gemmata* (Ach.) Massal.

A widespread species of moderately nutrient-rich bark on the trunks of mature trees in wayside and woodland situations, often common but rarely forming extensive colonies. Present on a wide range of trees but probably most frequent on *Fraxinus excelsior*. The species occurs throughout the British Isles but is much rarer in the Scottish Highlands and now absent from large areas of central England as a result of air pollution and the removal of suitable trees. Probably widespread in the less polluted parts of Europe, and perhaps circumboreal in the northern hemisphere, but its world distribution is in need of further investigation.

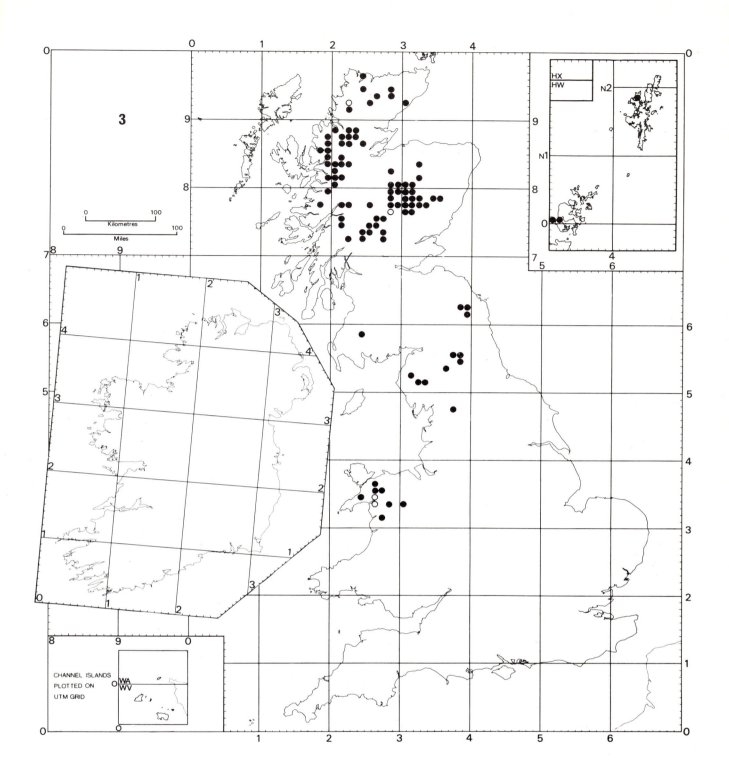

3 *Alectoria nigricans* (Ach.) Nyl.

An important and locally abundant component of arctic-alpine moss-lichen heaths, usually amongst mosses and ericaceous shrubs, especially in the Scottish Highlands, but also present at the higher altitudes of the Cheviots, Lake District, Pennines, Snowdonia and the Northern Isles; not known from Ireland. A distinctive species discolouring herbarium packets in a few years (see its separation from *Bryoria bicolor* under that species). There are no immediate threats to the status of the species. Outside the British Isles it has a circumpolar distribution in both northern and southern hemispheres and is arctic-alpine to northern boreal; interestingly, it is unknown from Africa, most of South America (except Tierra-del-Fuego) and the subantarctic islands.

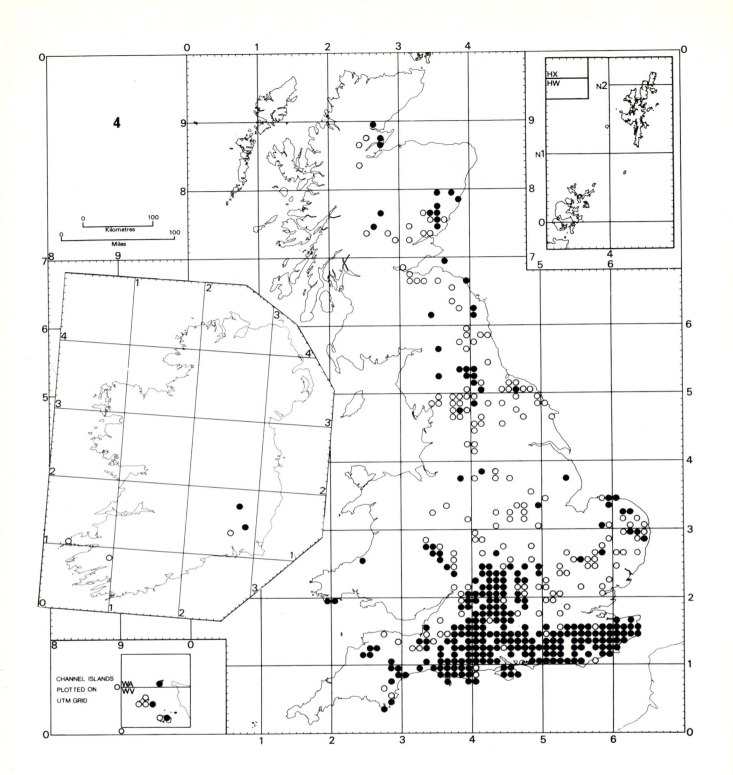

4   *Anaptychia ciliaris* (L.) Körber

Primarily a corticolous species of nutrient-rich, well-lit mature or old trees, especially *Fraxinus, Acer pseudoplatanus* and *Ulmus* species, on roadsides and in pastures influenced by nutrient-containing dust; often well-developed in nutrient-rich rain tracks, also rarely on similarly enriched rocks and tombstones. Markedly eastern in its distribution but extending westwards into low-rainfall (less than 95 cm p.a.) and sunny districts of Cardiganshire, Devon and Pembroke. An important component of the *Physcietum ascendentis* (*Parmelia acetabulum* facies). Formerly widespread, but unable to tolerate mean sulphur dioxide levels exceeding about 50 μg m^{-3}, and also inorganic fertilizer sprays; but tolerant, however, of automobile emissions, often occurring on trees by main roads. More recently (not signified by the map) its distribution has been much reduced by the large-scale loss of elm trees through Dutch Elm Disease. On maritime rocks in the west a narrow-lobed race, subsp. *mamillata* (Taylor) D. Hawksw. & P. James, also occurs. Widespread throughout southern Europe (on *Quercus* spp. in forests in the Mediterranean area), north Africa, extending into southern Scandinavia, Turkey, and the Canary Islands.

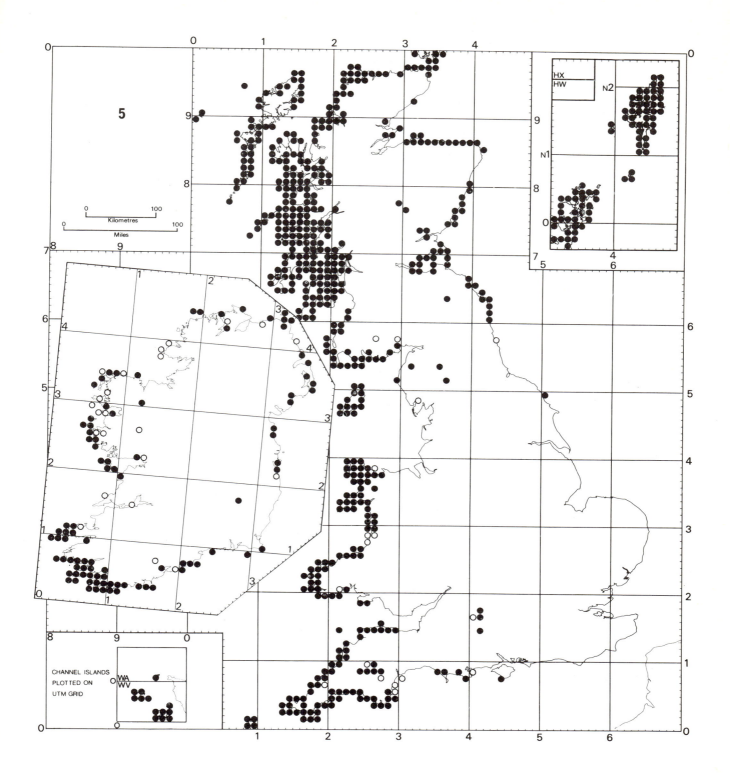

5 *Anaptychia fusca* (Huds.) Vainio

A xeric-supralittoral to terrestrial-halophilic species widely distributed and often very frequent on most rocky shores in south-west, west, north and north-east Britain; it occurs occasionally on trees where it enters the association *Teloschistetum flavicantis;* it is an important member of both the *Candelarielletum corallizae* associated with nutrient-enriched rocks and boulders, and the *Ramalinetum scopularis* characteristic of hard coastal rocks, and is found in a few inland sites as in Devon and Wiltshire (Stonehenge and Avebury) still subject to the influence of salt spray and local nutrient enrichment. Both shade and sun tolerant, it occurs on exposed or sheltered shores. Confined to Europe, distributed from north Norway and the Baltic to France, Spain and Portugal.

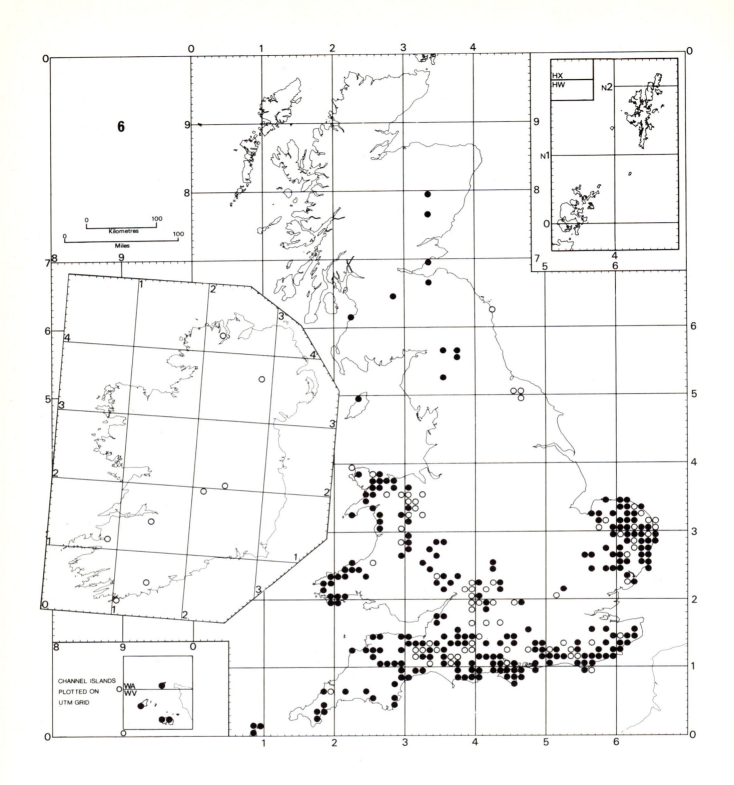

6   *Arthonia impolita* (Hoffm.) Borrer

This species is characteristic of the sheltered well-lit dry sides of usually ancient wayside, parkland and open woodland trees, particularly those with slightly nutrient-enriched bark such as *Acer pseudoplatanus* and *Ulmus* species. Often covering extensive areas, and the dominant species of the *Arthonietum impolitae*. Predominantly a southern species in the British Isles, with a few scattered outliers in northern England and Scotland, apparently requiring areas with high sunshine or low rainfall. Readily recognized by the pale mauve-grey slightly areolate thallus with a C+ rose reaction. Widespread in Europe, particularly in the west and south-west, and also reported from north Africa. Records from North America and elsewhere require reinvestigation.

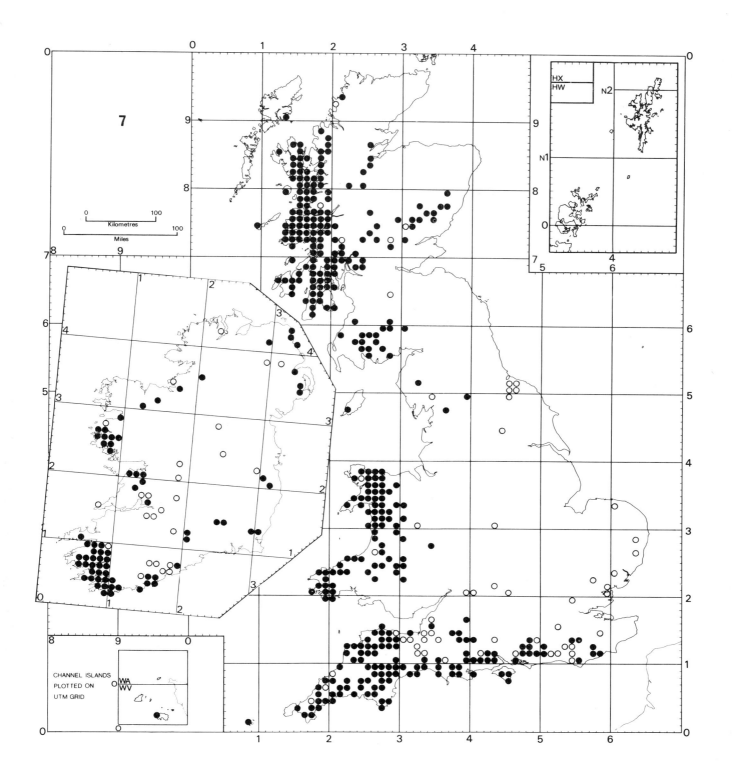

7 *Arthonia tumidula* (Ach.) Ach., s. lat.

A characteristic old woodland indicator species of smooth bark, especially *Corylus,* occurring both on twigs in the *Arthopyrenietum punctiformis,* and on smooth bark generally in shaded forest situations in the *Pyrenuletum nitidae;* less frequently on rough bark when in the latter association. Also allied to the *Graphidietum scriptae.* The species is now mainly southern and western, but the sparse older records suggest that it was probably present over most of lowland Britain; it is absent from large areas of the Scottish Highlands. Its reduction in area has probably been due to a combination of air pollution, lowering of water-tables, and clearance of established woodland. An aggregate of several closely related species varying with regard to the extent and colour of the white to pink or dark purple-red pruina covering the fruits. The aggregate is widespread in western, southern and central Europe, also in the Azores and Canary Islands and said to be widely distributed in the tropics and subtropics, but its occurrence is uncertain due to confusion with other species.

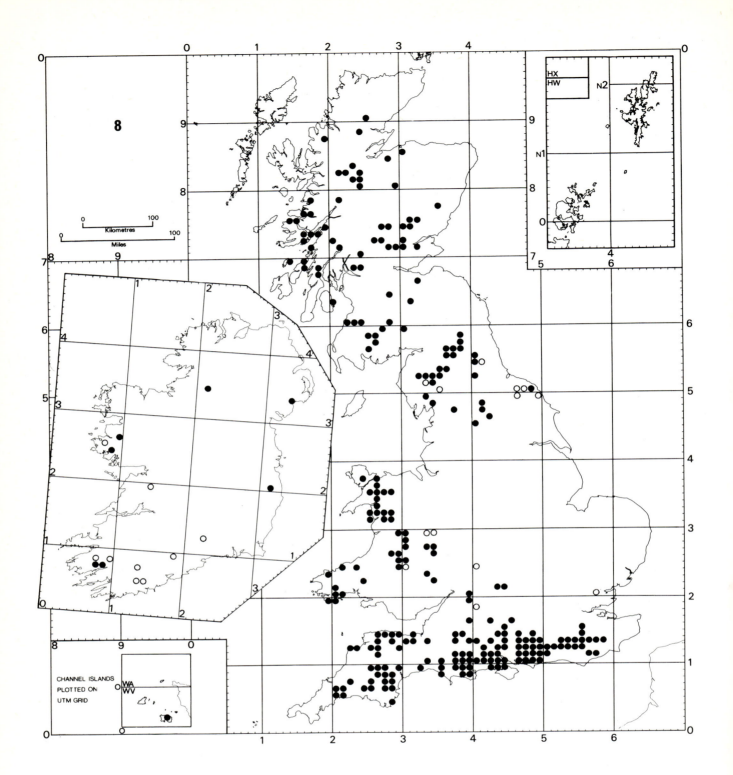

8   *Arthonia vinosa* Leighton

A species usually found on rough bark, almost exclusively on *Quercus,* in sheltered situations in woodlands with a long history of ecological continuity, and also on very ancient trees in more open situations in parklands. A member of the western facies of the *Lobarion pulmonariae* and the pre-*Lobarion* pioneer community, occurring on bare bark where there are breaks in the usual bryophyte cover of the *Lobarion*. Widespread in suitable habitats in the south and west of England, becoming more local, but still widely distributed, in northern England and Scotland. In Scotland it also occurs in communities akin to the *Calicion,* especially on old *Alnus*. Formerly, this species was probably widespread throughout central and eastern England but has been lost as a result of air pollution and the destruction of its habitat. Easily distinguished from allied species by the yellow-orange thallus (especially distinct around the fruits). Distributed throughout Europe where relics of the *Lobarion* still occur; extra-European distribution uncertain.

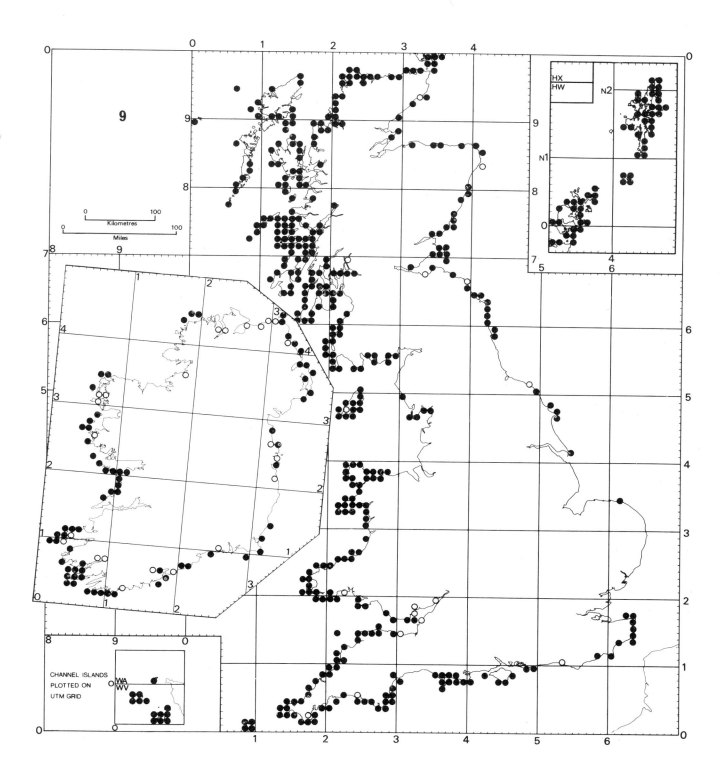

9 *Arthopyrenia halodytes* (Nyl.) Arnold, s. lat.

On all marine rocky substrates, from calcareous to siliceous, including chalk, shingle, cement and brickwork; also on the shells of many littoral invertebrates. A strictly littoral lichen, extending from low water neap tide levels to the splash zone. It often penetrates upshore in brackish water runnels. The dominant lichen in the littoral zone of moderately to very wave-exposed rocks, it is invariably fertile. Abundant throughout Britain, but largely absent from the east and parts of the south, where muddy substrates predominate. It is the most pollution-tolerant marine lichen and penetrates well into estuaries where the silt load is appreciable. Possibly no change of distributional status except where marine silt-load has increased. It has been much confused with *A. sublittoralis* but is much commoner, and has been overlooked as an inconspicuous brown stain on rocks. Probably cosmopolitan at least in the northern hemisphere, from arctic to warm temperate waters; it is rare in the tropics. Its world ecology is similar to that in Britain but it becomes rare and confined to shade and crevices in warmer climates.

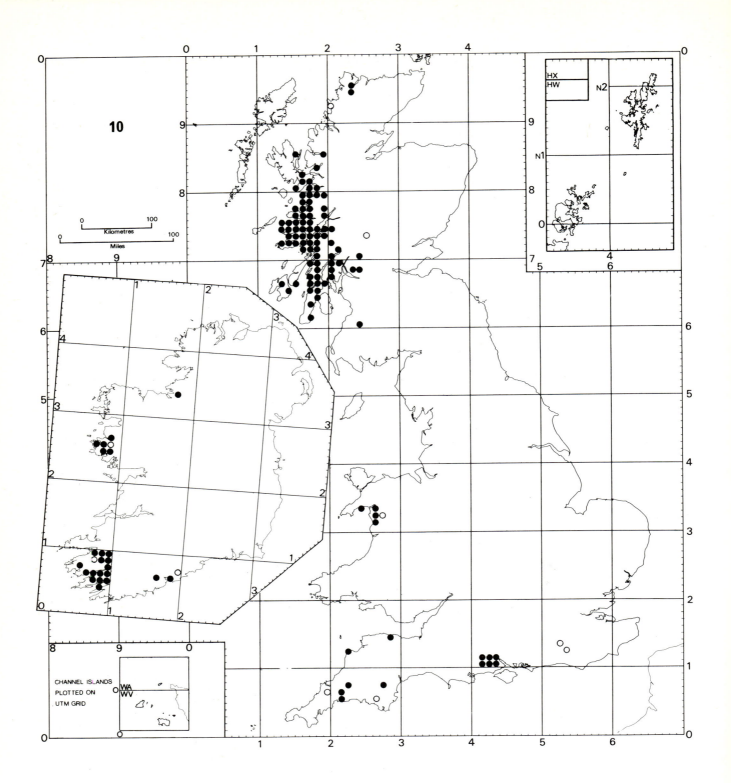

10　*Arthothelium ilicinum* (Taylor) P. James

Primarily an oceanic species of high rainfall areas, now confined to the New Forest and a few undisturbed ancient woodland sites in south-west England and Wales, but widespread and often locally abundant in south-west and western Ireland and western Scotland. In England and Wales it is largely restricted to old *Ilex aquifolium* trunks in sheltered humid situations, but in Scotland and Ireland it occurs on the smooth bark of a wide range of trees *(Ilex, Corylus, Fraxinus, Fagus* and *Sorbus).* It is a member of, or allied to, the *Graphidietum scriptae.* Presumably formerly widespread in woodlands with a long history of ecological continuity, but now restricted due to their destruction and its apparently limited powers of dispersal. The species occurs at least in the Azores, Canary Islands, western France (Brittany) and Portugal, and is also known from the west coast of North America and Norway. The British Isles and Norway represent the most northerly localities in Europe.

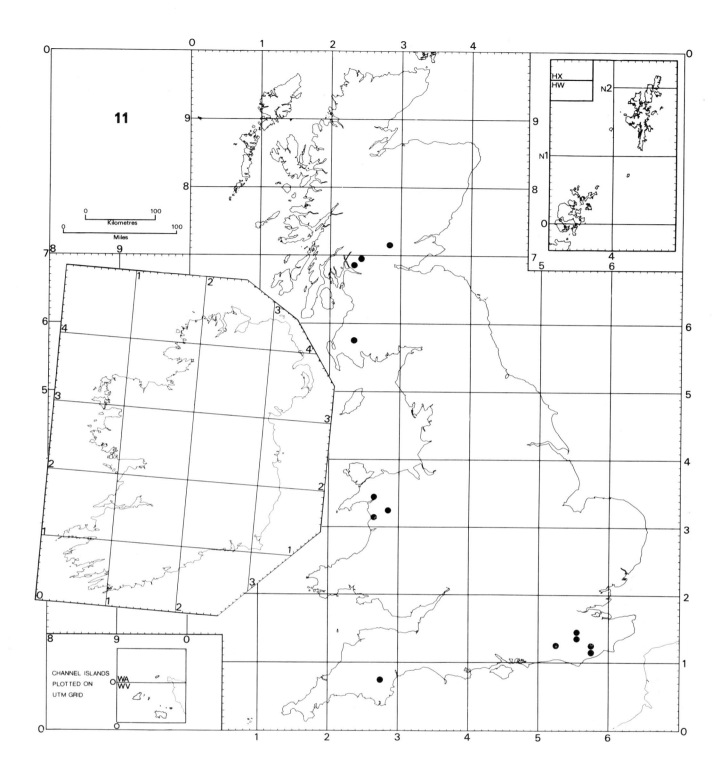

11   *Arthothelium ruanum* (Massal.) Zwackh

A corticolous species, rare and local in southern England, mid-west Wales and lowland central and south-west Scotland. On smooth bark, especially young *Fraxinus* (also *Castanea, Corylus* and *Sorbus*), in sheltered situations (usually by streams) in old, but often previously managed woodlands. A principal member of the *Opegraphetum herpeticae*. Incorrectly reported from Ireland. Also found in southern Scandinavia and central Europe; possibly present in north-east America.

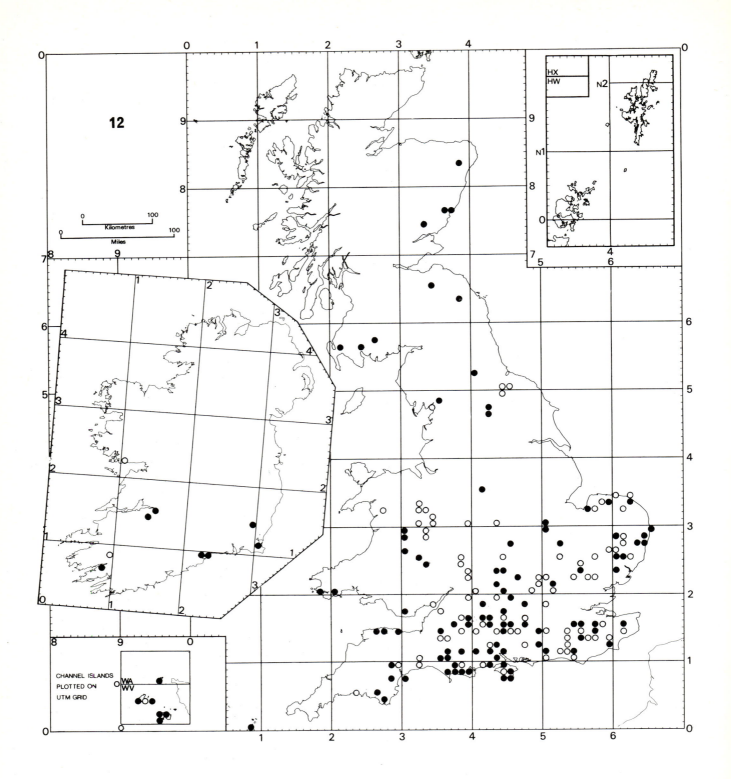

12 *Bacidia incompta* (Borrer ex Hooker) Anzi

This species is almost exclusive to *Ulmus* bark. It is often associated with exudates from wounds where it may form extensive dark green subgelatinous colonies, often in association with *Caloplaca luteoalba*. Especially characteristic of old wayside and parkland trees, in open situations with a low rainfall, in the southern and eastern parts of the British Isles, becoming very rare in Scotland. It has been severely affected by air pollution and the felling of elms; the map presented here shows the distribution prior to the ravages of Dutch Elm Disease, and it has almost certainly disappeared from many of the localities shown. However, it also occurs in wound-tracks on old *Fagus*, especially in the New Forest. It is known from southern Scandinavia and much of temperate Europe; also from North America.

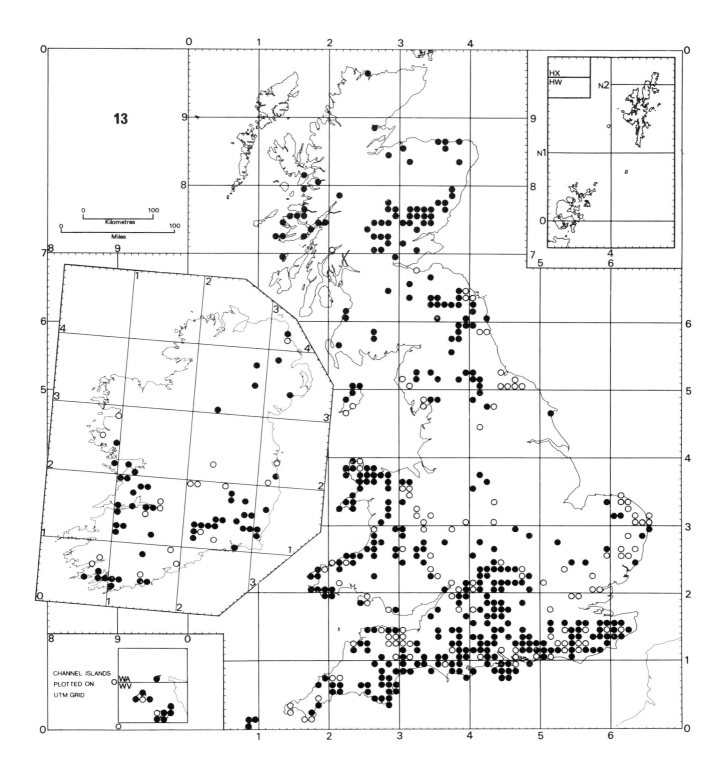

13 *Bacidia rubella* (Hoffm.) Massal.

A species characteristic of moderately nutrient-rich bark of broad-leaved trees in wayside, parkland and open woodland situations. Most frequently found on *Acer pseudoplatanus, Fraxinus excelsior, Sambucus nigra* and *Ulmus* species in the *Gyalectetum carneolutae, Physcietum ascendentis* and communities intermediate with the *Lobarion pulmonariae*. Sometimes without apothecia and then easily overlooked. The rather similar *B. phacodes* is distinguished by the shorter ascospores and smoother thallus. Formerly widespread in lowland Britain but becoming rarer in Scotland (almost entirely absent from the Scottish Highlands). Now scarce to absent from large areas of central and eastern England due to a combination of air pollution, the felling of wayside trees, and agricultural sprays; absent from areas with a mean sulphur dioxide level exceeding about 50 μg m^{-3}. The species has undoubtedly declined more recently than indicated by this map due to the effects of Dutch Elm Disease. Widespread in western and south-western Europe, and particularly common, even in oak forests, in southern France and northern Italy, but absent from northern boreal regions.

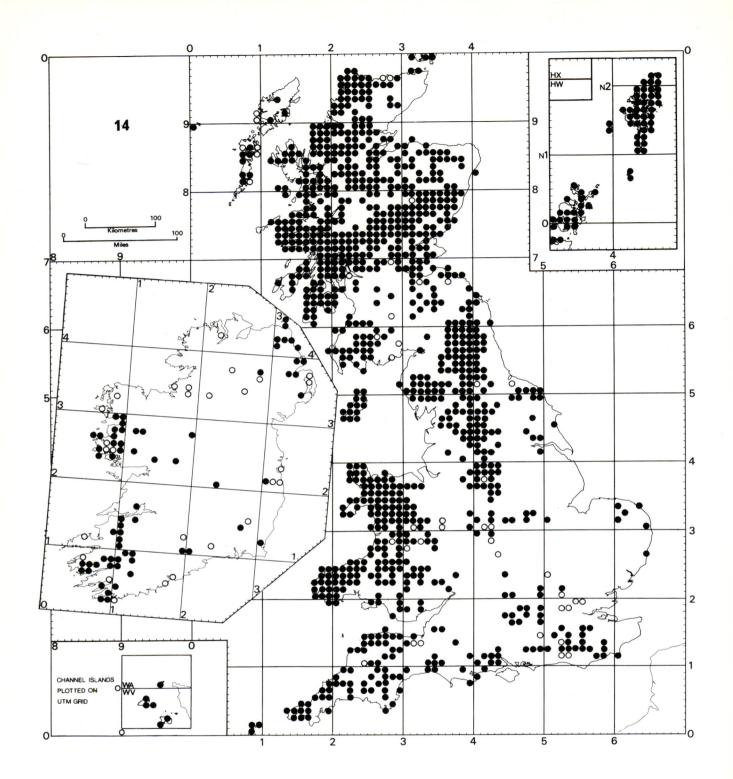

14 *Baeomyces rufus* (Huds.) Rebent.

A widespread species in heathland and well-lit acid woodland habitats where it often forms extensive colonies over-growing consolidated soils, especially on sloping banks and old paths, and also on water-retentive siliceous rocks and then often in sheltered situations or by streams. Also noted rarely on the bases of shaded tombstones. It extends from coastal and lowland to upland sites; its apparent scarcity in central and eastern England is largely due to the disappearance and current rarity of suitable habitats. The species is frequently sterile and then easily overlooked. *B. roseus* is readily separated by the grey-white thallus with scattered white nodules, and when fertile, by the bright rose-pink (not brown) stalked apothecia. This species is an early coloniser of burnt heathland and it is only likely to be affected by the destruction of heathlands generally. A common circumboreal species in the northern hemisphere; also present at least in the Azores and Canary Islands.

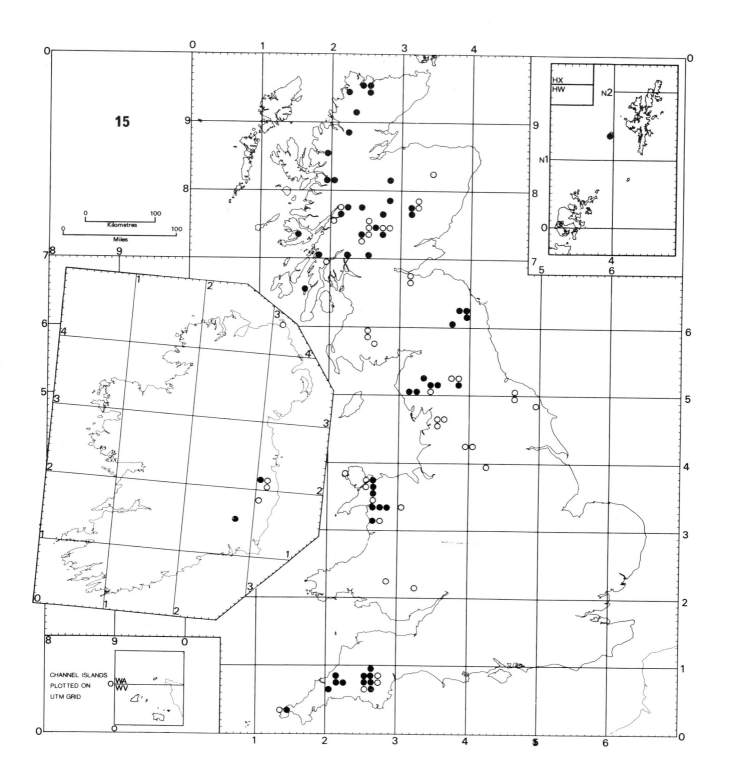

15 *Bryoria bicolor* (Ehrh.) Brodo & D. Hawksw.

A local species of suboceanic upland areas, occurring mainly amongst mosses but sometimes directly on soil or rock. Distinguished from *B. smithii* (Du Rietz) Brodo & D. Hawksw. by the more tufted and delicate habit and the consistently PD + red reaction (at least in parts), and from *Alectoria nigricans* by the small size, lack of conspicuous pseudocyphellae, presence of lateral short perpendicular spinulose branches, and lack of reaction with K and C. The species has been lost from parts of the Pennines and South Wales probably due to a combination of the effects of air pollution and heathland management. A sub-oceanic montane species in Europe, though now scarce; also known from North America, Himalaya, Japan, Malaysia and east Africa.

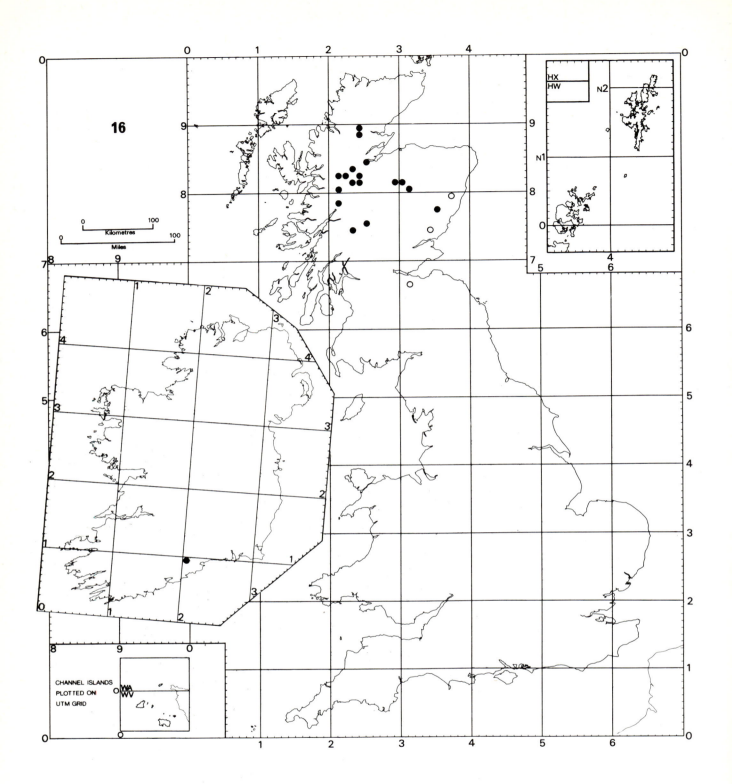

16 *Bryoria capillaris* (Ach.) Brodo & D. Hawksw.

A rather rare, but locally plentiful, species of ancient pinewoods in central and eastern Scotland with an old record (unlocalised) from Yorkshire and one recent report from Ireland. Mainly on conifers and *Betula* on which it forms part of a distinctive community, the *Usneetum filipendulae*. Distinguished from *B. subcana* (Nyl. ex Stizenb.) Brodo & D. Hawksw. by the intense K+ bright yellow, KC+ rose reaction of the cortex as well as the generally more slender habit; herbarium packets containing this species become discoloured reddish-brown in a few years. The species is likely to be adversely affected by any depletion of the native pinewoods of the British Isles, but it is also sensitive to sulphur dioxide and fluoride pollution. A widespread, southern-boreal to north temperate (hemiboreal) species; also known from central Europe and Scandinavia, the Pyrenees, North America, the Canary Islands and Japan; probably circumboreal.

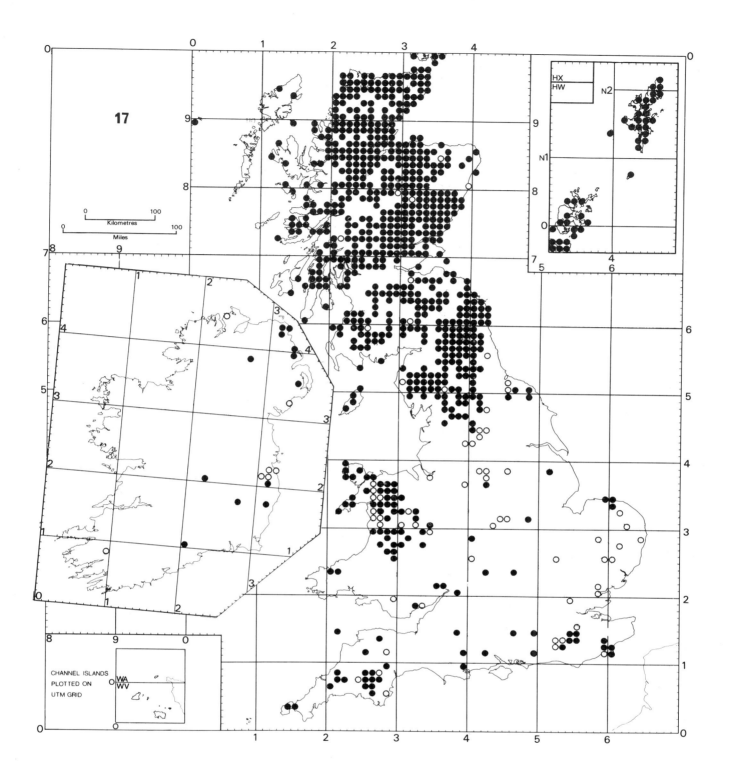

17 *Bryoria fuscescens* (Gyelnik) Brodo & D. Hawksw.

A characteristic species of the *Pseudevernietum furfuraceae,* widespread in upland, northern and eastern parts of the British Isles but becoming scarcer in extremely oceanic areas. Particularly characteristic of coniferous forests in the north but also occurring on siliceous rocks, amongst mosses, on walls, and, especially in lowland England, on fence rails. A very variable species becoming contorted and spinulose in exposed situations, but generally easily recognisable by the tendency to have a pale base, fissure-like as well as tubercule-like soralia, and the matt smoky-brown colour; the soralia are always PD+ red, but the thallus is PD+ or PD−. Moderately sensitive to sulphur dioxide pollution and absent where this exceeds about 55 μg m^{-3}, the species consequently declined in many parts of lowland Britain, but has started to expand again in the last few years as evidenced by its recent discovery as young plants in Buckinghamshire and Surrey. Widely distributed in Europe and North America, especially outside old established boreal forest areas and with a lowland tendency in these regions; probably circumboreal; also present in the Canary Islands and east Africa.

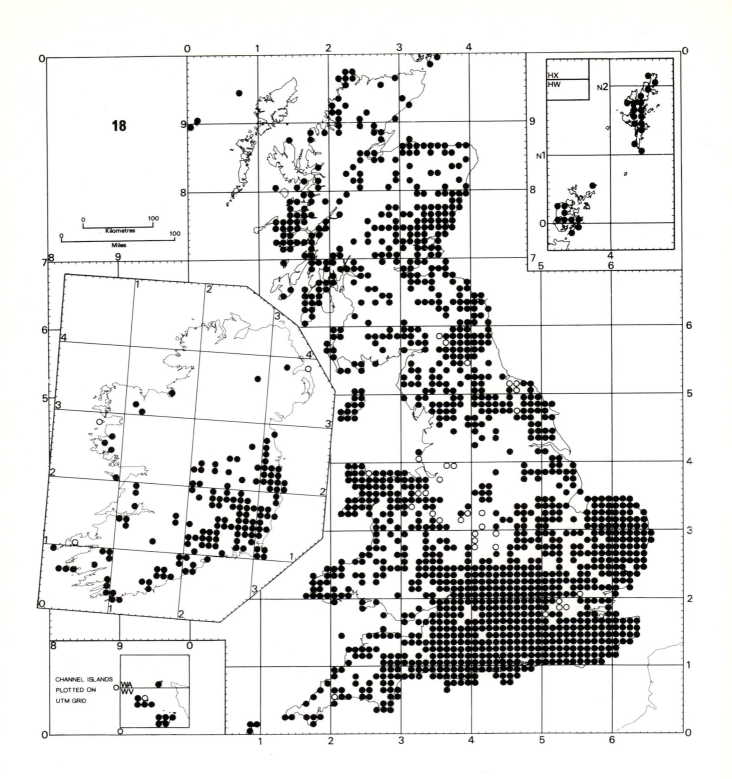

18 *Buellia punctata* (Hoffm.) Massal.

A corticolous and saxicolous species of wide ecological amplitude and tolerant of moderately high pollution by SO_2 (up to 100 μg m^{-3}) and inorganic fertilizers. Enters into the following associations: the *Arthonietum impolitae* on somewhat dry basic-barked trees; the *Buellietum punctiformis* on nutrient-enriched or hypertrophicated bark relatively insensitive to pollution; the *Physcietum ascendentis* on nutrient-rich bark in unpolluted sites; the *Candelarielletum corallizae* on ornithocoprophilous rocks; and occasionally in the *Ramalinetum scopularis*. It is a widespread, often common species over much of Britain, absent only from the most severely polluted areas of England and upland areas in Scotland. It is probably under-recorded. Widespread throughout the north-temperate zone, and also recorded from several countries in the southern hemisphere.

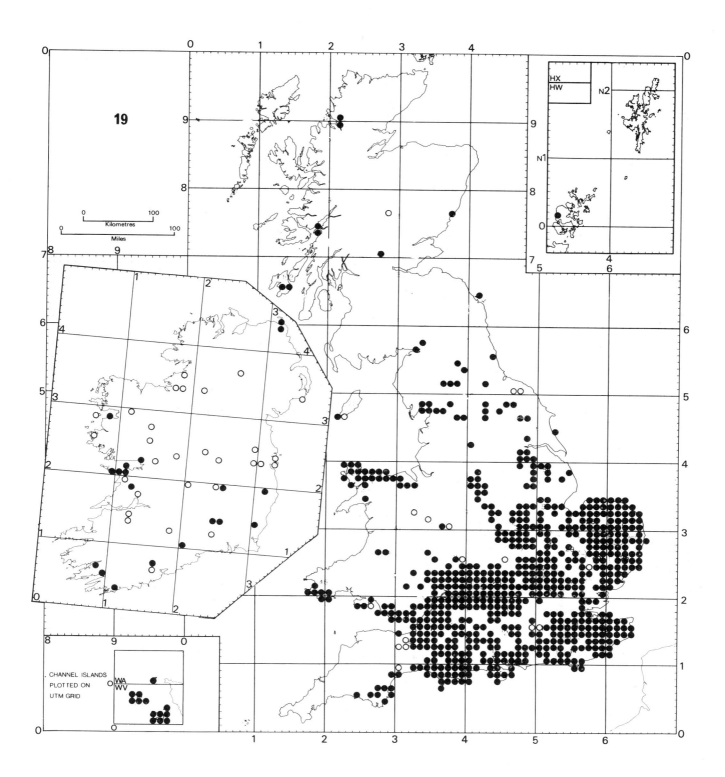

19 *Caloplaca aurantia* (Pers.) Hellbom

Like *C. heppiana* (map 21), *C. aurantia* is markedly south-eastern in distribution in Britain and is less widely distributed west of a line drawn from the Severn Estuary to the Wash. To the east of this line the species is common on man-made calcareous substrates chiefly in churchyards and occasionally on old walls and asbestos-cement roofs; west of the line *C. aurantia* is most frequent on well-lit, dry exposures of limestone as in Derbyshire, North Wales, Pembroke and the Island of Lismore. It is pollution-tolerant, occurring in areas with up to 100 μg m^{-3} SO$_2$. It forms an important element of the eastern facies of the *Caloplacetum heppianae*. It is readily distinguished from *C. heppiana* by its flattened, often zoned, contiguous lobes. Widespread, at least in western Europe, with a marked southern tendency.

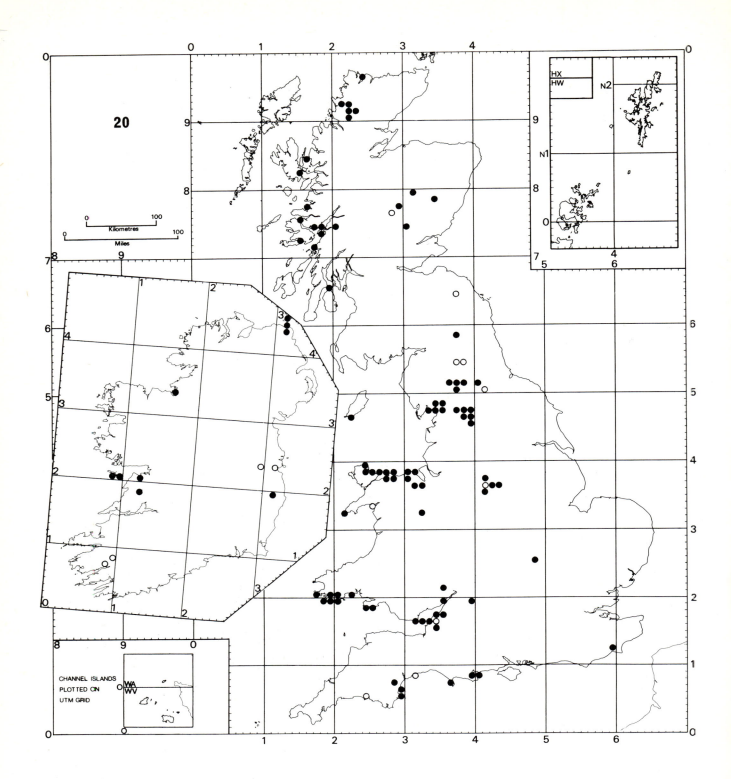

20 *Caloplaca cirrochroa* (Ach.) Th.Fr.

A widespread but local species almost entirely confined to pure, hard limestones. It belongs to the *Gyalectetum jenensis* and grows on sheltered, rather damp, underhangs or vertical rock faces often partly obscured by summer vegetation or shrubs. It is widely distributed in England, but very rare in the south (where it also occurs on ancient buildings). It is probably widespread but overlooked in Ireland. It occurs on limestone in central Europe from northern Italy to southern Scandinavia. Reliably recorded from South Dakota, and in need of confirmation from several other parts of North America.

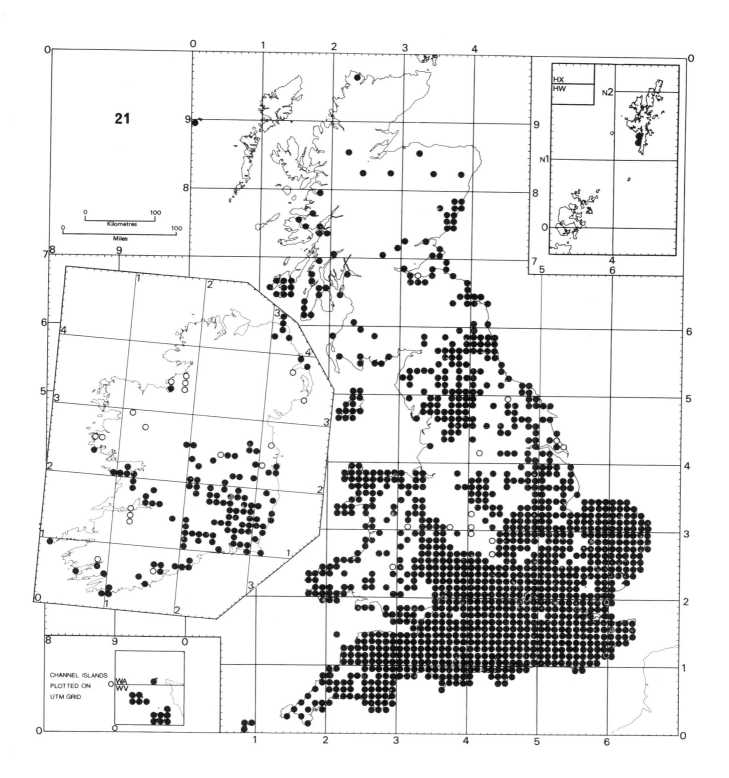

21 *Caloplaca heppiana* (Müll. Arg.) Zahlbr.

This species has a markedly south-eastern distribution in Britain, but is more widely distributed, in areas west of a line drawn from the Severn Estuary to the Wash, than *C. aurantia*. In eastern England it is very widespread and often abundant on man-made calcareous substrates such as gravestones, mortar, stone or brick walls and asbestos-cement roofing. It is also pollution-tolerant, surviving concentrations of up to 200 μg m^{-3} SO$_2$. In western areas it also grows on natural limestones, including oolite, and seems to prefer slightly more sheltered sites than *C. aurantia* with which it often grows. Some of the coastal records may be errors of identification for *C. thallincola. C. heppiana* is the dominant species of most facets of the *Caloplacetum heppianae;* the fact that it is also frequently well represented in the *Placynthietum nigri, Gyalectetum jenensis* and *Physcietum caesiae* associations is due to its wide ecological amplitude. Very widespread in Europe, with a marked southern tendency, and more often on man-made substrates than *C. aurantia*.

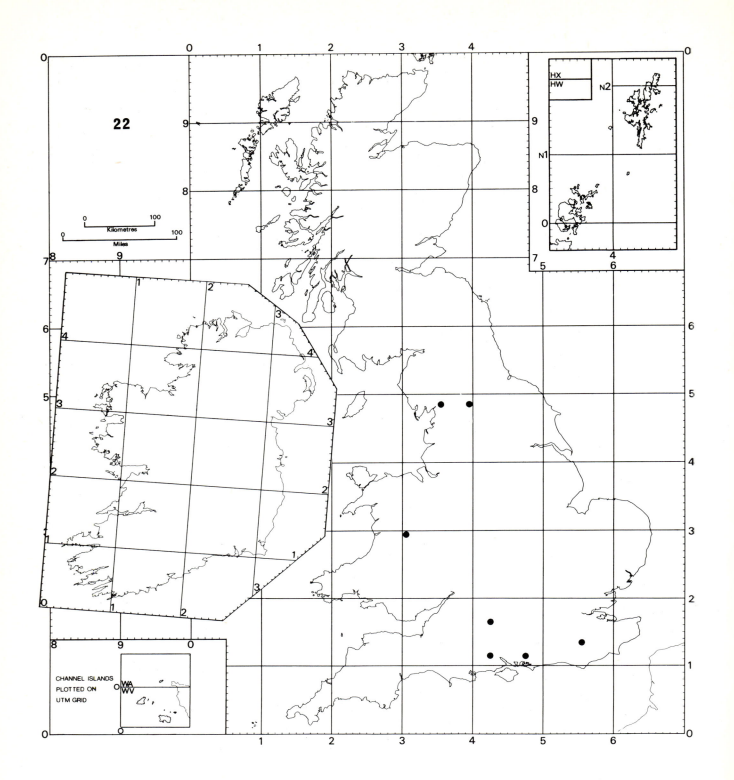

22 *Caloplaca herbidella* (Hue) Magnusson

A distinctive isidiate species characteristic of the trunks of ancient *Quercus* trees in sites with a long history of ecological continuity, occurring in well-lit situations in *Pertusaria*-rich facies of the *Parmelietum revolutae*. Extremely rare and local in the British Isles, often on only a single tree in a site, the stations widely scattered and probably relict. Evidently absent from high rainfall areas of the British Isles, and with a strong southern continental distribution in Europe. Recorded also from Austria, Bulgaria, Denmark, France, Germany, Hungary, Italy, Yugoslavia, Sweden and Switzerland.

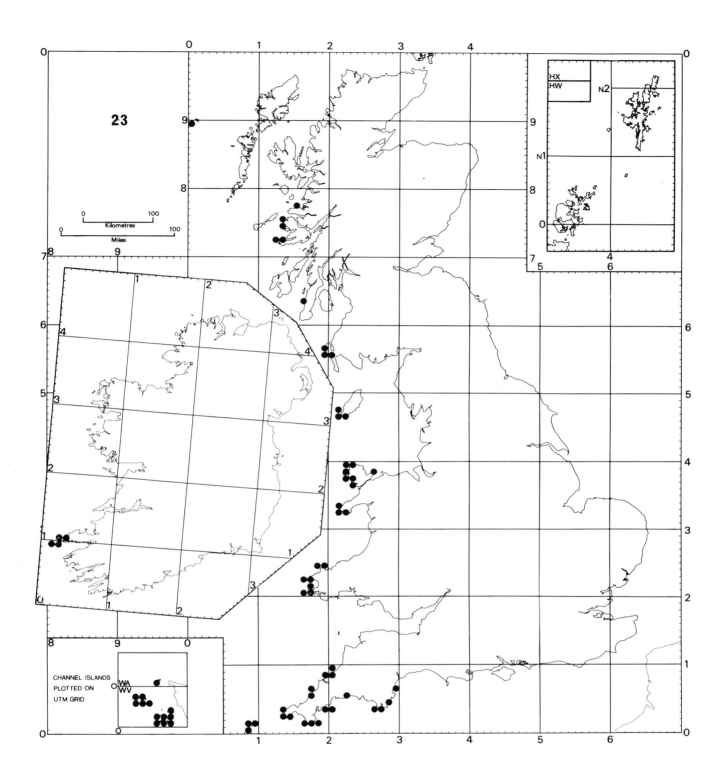

23 *Caloplaca littorea* Tavares

Found on siliceous rocks in general, slates and mudstones. Locally common in west Britain, especially on islands and peninsulas, usually on seashores in the mesic and xeric supralittoral zones. It is mostly confined to crevices and below sheltered overhangs, but in places appreciably exposed to winds and sunshine, and often on soft, eroding, dusty rocks. It rarely forms extensive patches, and is seldom fertile. It is presumably intolerant of excessive wetness or weathering and may prefer warm, well-lit, nutrient-enriched sites. No change in distribution can be discerned since it was first recognised in Britain in 1961. It has probably been much overlooked by recorders and has been confused with forms of *C. marina* and *C. verruculifera;* west Scotland and south and west Ireland are certainly under-recorded. Distributed in south-west Europe from Scotland to the Mediterranean, where its ecological preferences need not be exclusively maritime.

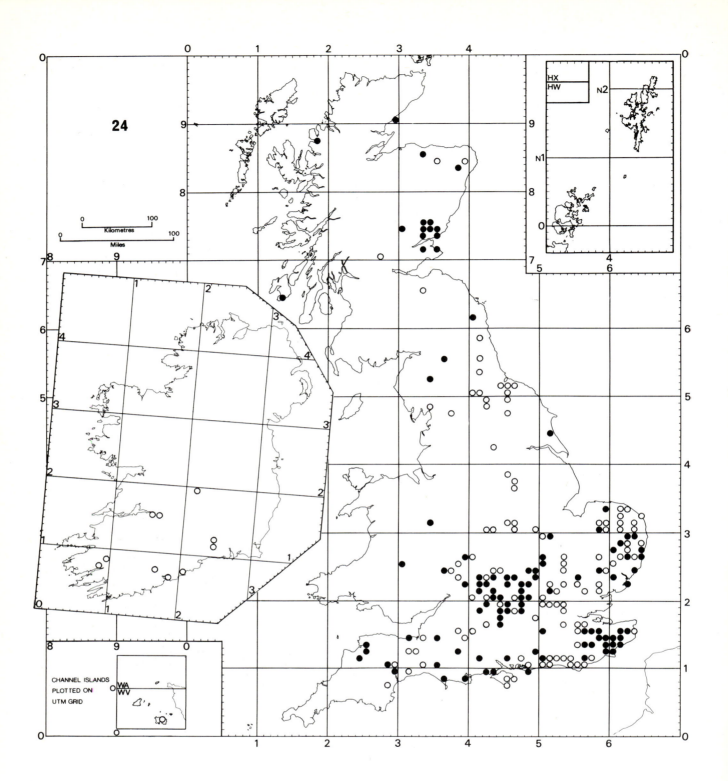

24 *Caloplaca luteoalba* (Turner) Th. Fr.

A corticolous, very rarely saxicolous, species almost always associated with the nutrient-rich exudates from wounds on the boles of *Ulmus,* rarely *Acer pseudoplatanus* and *A. campestre.* Frequently associated with *Bacidia incompta* (map 12), it forms a special facies of the *Physcietum ascendentis.* Now very rare; before Dutch Elm Disease *C. luteoalba* had a scattered, predominantly eastern distribution in Britain and was a characteristic species of old wayside and parkland *Ulmus.* Previously much reduced in distribution by pollution from inorganic fertilizers, the species must now have been almost exterminated, except for a few sites on *Acer* spp., throughout Britain. The map shows the distribution before the onset of Dutch Elm Disease. Widespread in western Europe on similar substrates, but beginning to decline as Dutch Elm Disease spreads.

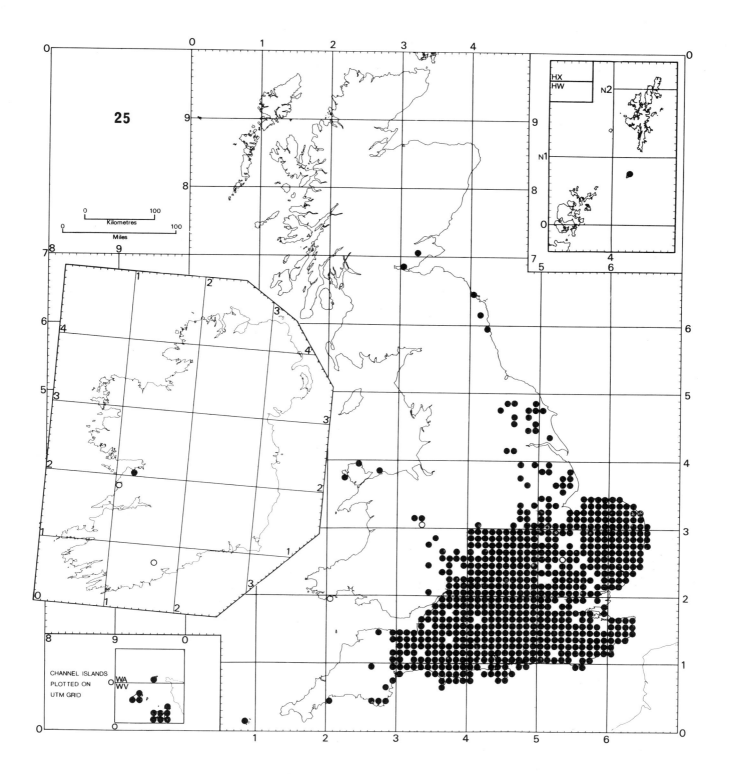

25 *Caloplaca teicholyta* (Ach.) Steiner

A widespread, saxicolous species often abundant in south-east England, but becoming rare west and north of a line between Exeter, the Severn Estuary and the Wash. Characteristic of markedly basic man-made substrates such as gravestones, stone-mortar walls and asbestos-cement roofing. Generally sterile, but occasionally fertile on south-facing church parapets and window ledges and on natural rock outcrops such as the outliers in Fair Isle, Anglesey, the Great Orme and the Isles of Scilly. *C. teicholyta* belongs to the eastern element of the *Caloplacetum heppianae* which is characteristic of dry, sunny, calcareous surfaces. Abundant in southern and central Europe, but limits of distribution poorly understood.

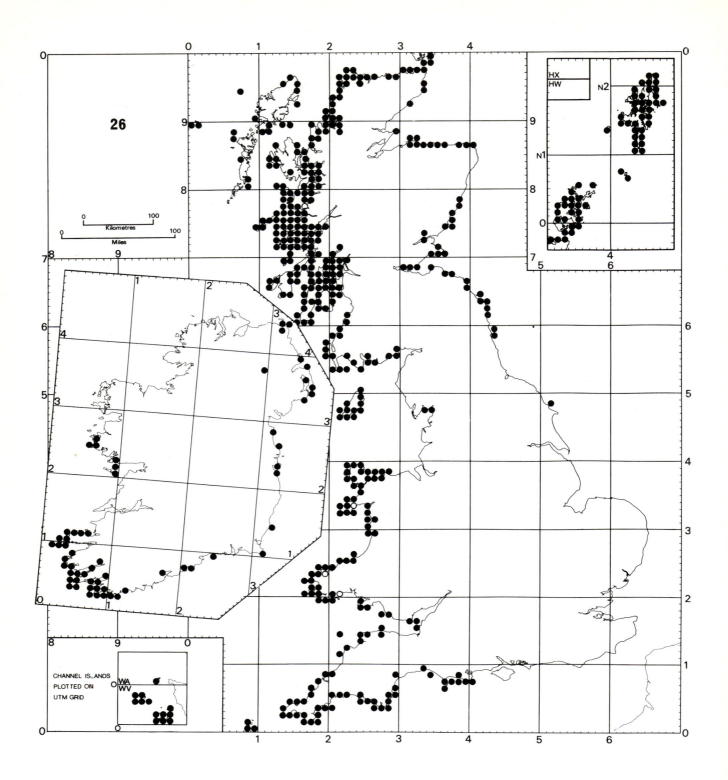

26 *Caloplaca thallincola* (Wedd.) Du Rietz

Found on both siliceous and calcareous rocks. Very common where hard rocks, subject to little weathering, are found. Absent from most of east and south-east England owing to muddy or soft substrata. A plant of the seashore in the mesic and submesic supralittoral zones; intolerant of immersion in seawater but requiring periodic splash and spray. It is usually best developed on shaded shores, below overhangs, or on north and east facing sites. Usually fertile; it is intolerant of marine pollution. No change of status is discernible. The plant is well recorded except in east Scotland, the Outer Hebrides, Cumbria and Ireland in general. It is probably much confused with *C. heppiana* when on calcareous shores, though the latter plant is never mesic or submesic-supralittoral in distribution. Abundant in north-west Europe with an ecology similar to that in Britain.

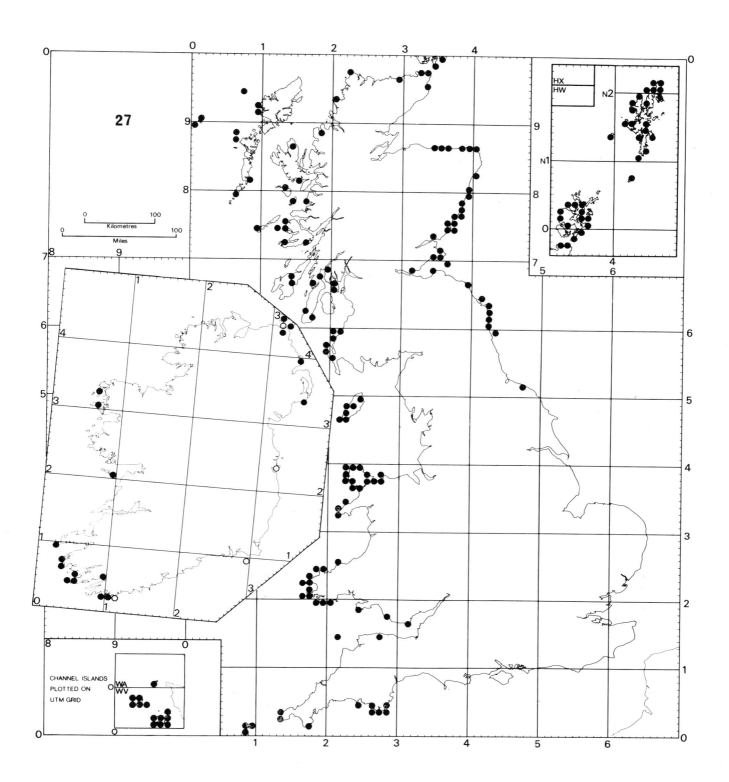

27 *Caloplaca verruculifera* (Vainio) Zahlbr.

An exclusively maritime species which is locally abundant on nutrient-enriched acid, and sometimes basic, rocks often on or near sea bird colonies or birds' perching stones. Its position on the shore is mesic-supralittoral to terrestrial and, with *Aspicilia leprosescens, Lecanora poliophaea* and the alga *Prasiola quadrata,* comprises a nutrient-enriched facies of the *Caloplacetum marinae.* The species is confined to more or less horizontal rock faces and develops best in rather shaded, sheltered habitats where it is often fertile. It is widely distributed on all rocky shores in west and north Britain, being particularly abundant on off-shore islands (e.g. Skomer), Pembroke and the south Devon coastal schists. Essentially a northern maritime species in Europe common around the Scandinavian coast and extending to Spitsbergen, Novaya Zemlya and Jan Mayen. Distribution elsewhere uncertain due to confusion with other species.

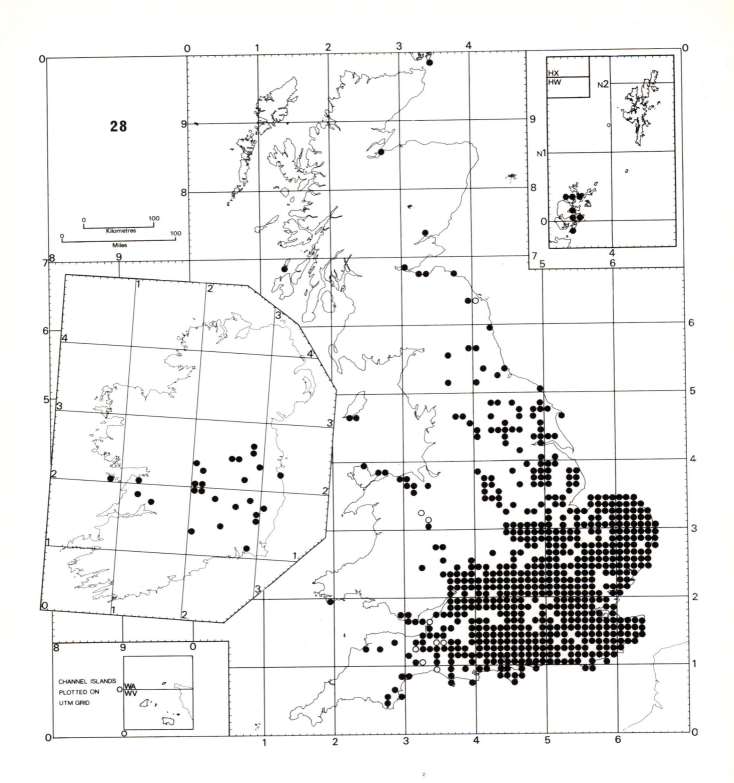

28   *Candelariella medians* (Nyl.) A. L. Sm.

The species has a predominantly eastern distribution in Britain, mainly concentrated to the east of a line drawn between Exeter, the Severn Estuary and the Humber estuary, Yorkshire. It is especially abundant on calcareous man-made materials, especially gravestones, where it enters the associations *Caloplacetum heppianae* and, particularly, the nutrient-enriched *Physcietum caesiae*. *C. medians* prefers well-lit sites, especially those enriched by bird droppings; the tops of gravestones are preferred. It is very rarely recorded on natural substrates. It is pollution tolerant, occurring in areas with up to 100 μg m^{-3} SO$_2$. It is widespread in western Europe, north to south Scandinavia, in similar places.

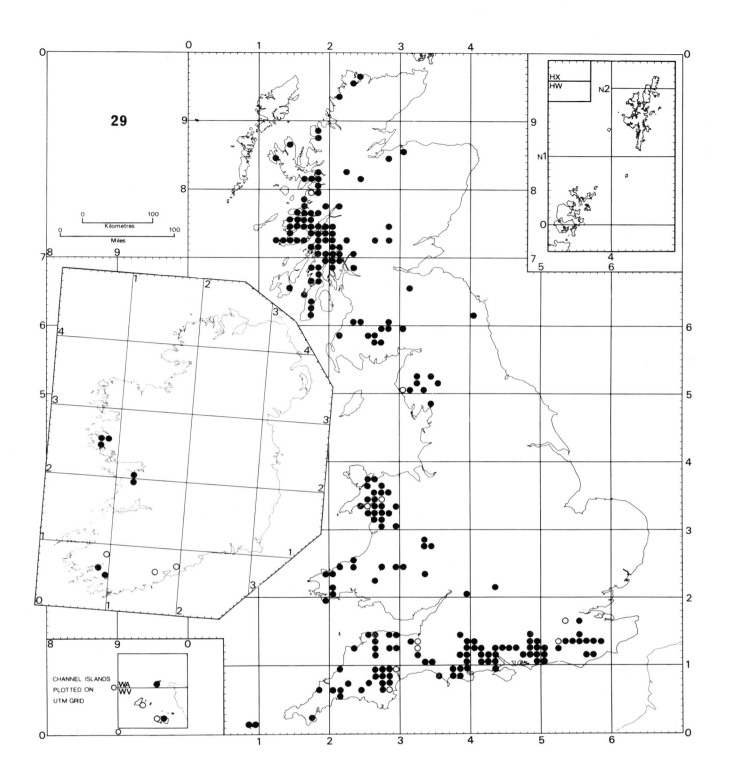

29 *Catillaria atropurpurea* (Schaerer) Th.Fr.

A species of the *Lobarion pulmonariae* represented in both its typical western European facies as well as in the pre-*Lobarion* nodum, the latter characteristic of many parts of southern England. It most frequently colonizes mossy bark of the boles of a diverse range of phorophytes, *Fraxinus, Quercus, Salix* and *Ulmus* being the most usual substrates. *C. atropurpurea* is tolerant of deeply shaded as well as well-lit situations, occurring in sheltered, moist, humid ravines as well as more open woodland, usually on rather rich soils. In the south of England it is mainly an old woodland relict species; in Scotland it is more generally distributed, in the west largely confined to woodland, but sometimes also occurring on wayside trees by streams or bogs. Abroad it is an oceanic species known from western Norway, France, the Iberian peninsula, the Azores and the Canary Islands. It probably occurs in most oceanic temperate areas of the world and is definitely known from Tasmania and New Zealand.

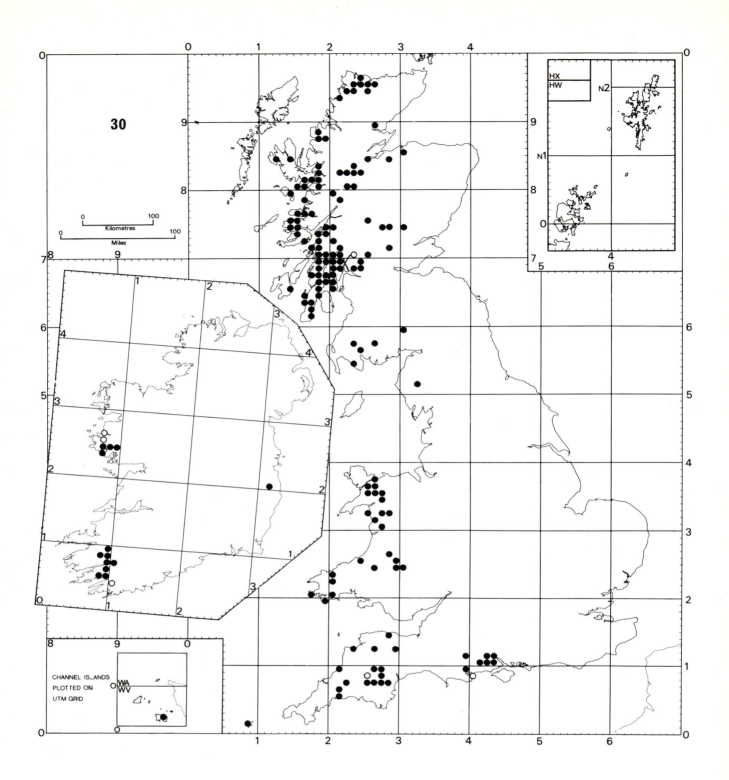

30   *Catillaria pulverea* (Borrer) Lettau

C. pulverea occurs in sheltered, moist, often ill-drained carr and damp scrub woodland. It is often present in boggy clearings in old woods or on thinly dispersed trees in wet places near streams and rivers. It grows on smooth to moderately rough, probably more or less leached, bark of *Salix, Corylus, Fraxinus, Quercus, Sorbus aucuparia, Betula* and *Alnus*. Widespread, often locally frequent in western Scotland, it belongs to the *Parmelietum revolutae* and an interrelated species-poor facies of the *Lobarion pulmonariae*. On the other hand, in carrs, especially in England and Wales, it strays into a shaded facies of the *Lecanoretum subfuscae*. Although widespread in western Scotland, and probably western Ireland, *C. pulverea* is more restricted in England and Wales due to scarcity of suitable sites. The species is frequently sterile and thus may be under-recorded on the map. Abroad, it is known from western Scandinavia and France; a fertile, non-sorediate counterpart *Catinaria albocincta* occurs in the Azores.

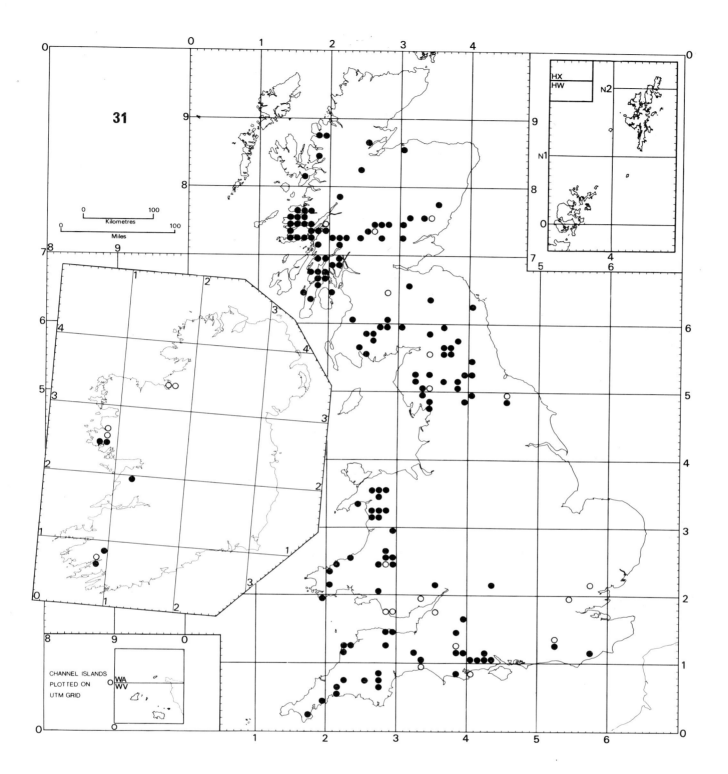

31 *Catillaria sphaeroides* (Dickson) Schuler

A species of the typical western European facies of the *Lobarion pulmonariae*. It occurs most frequently on mossy bark at or near the bases of old trees, particularly *Fraxinus* and *Quercus* in damp, often very sheltered, humid, very shaded situations; it also colonizes mossy boulders in similar sites. It is essentially a valley species and in the south of England it is an indicator species of ancient woodland, especially in tree-lined ravines. In northern England and Scotland *C. sphaeroides* is more widespread, but nevertheless is confined to shaded woodland, sometimes near water-falls, in steep-sided valleys, or in moist woodland on rich alluvial or basic volcanic soils. Distribution elsewhere uncertain due to confusion with other species; present at least in Brittany and Denmark where it is rare.

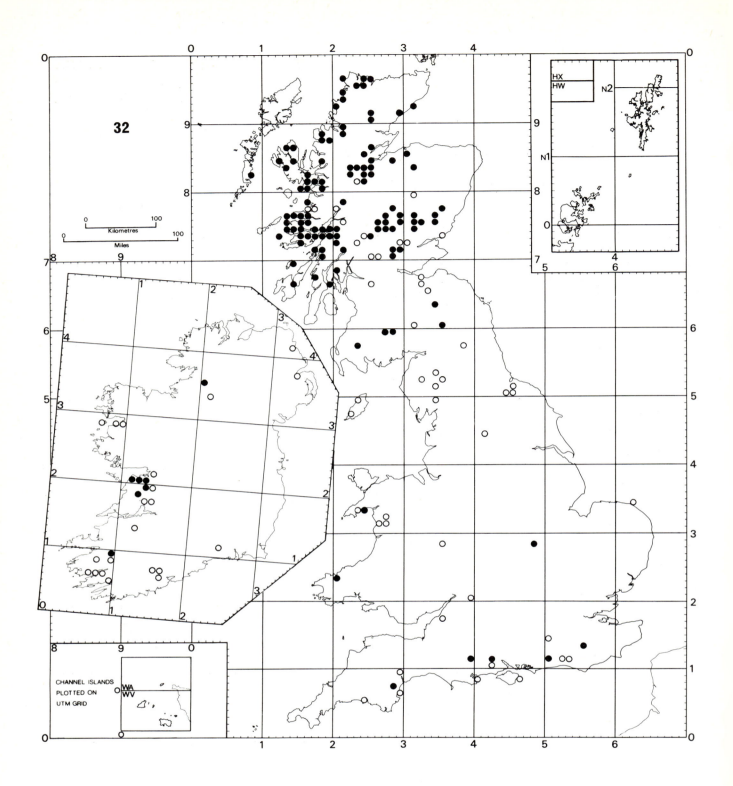

32   *Catinaria grossa* (Pers. ex Nyl.) Vainio

A corticolous species of old trees with more or less nutrient-rich, basic bark (pH 4.9-5.8) in well-lit to very shaded situations in rather humid woodland, or, more rarely, wayside sites in boggy or sheltered valleys, on basic or rich alluvial soils. It is more frequent on *Fraxinus,* but also occurs on *Ulmus, Corylus, Quercus* and *Salix.* It is a member of a moisture-loving facies of the pre-*Lobarion* community as well as the *Lobarion pulmonariae.* The species is still widespread, if rather local, in Scotland occurring in a facies of the *Lobarion pulmonariae* rich in cyanophilous species as well as in more sheltered lowland areas of the central Highlands. In England, where each dot on the map represents a single tree, it has much decreased, probably due to air pollution, forest management and drainage. It is widespread and probably common in most areas of Ireland. Abroad, it is a northern oceanic species.

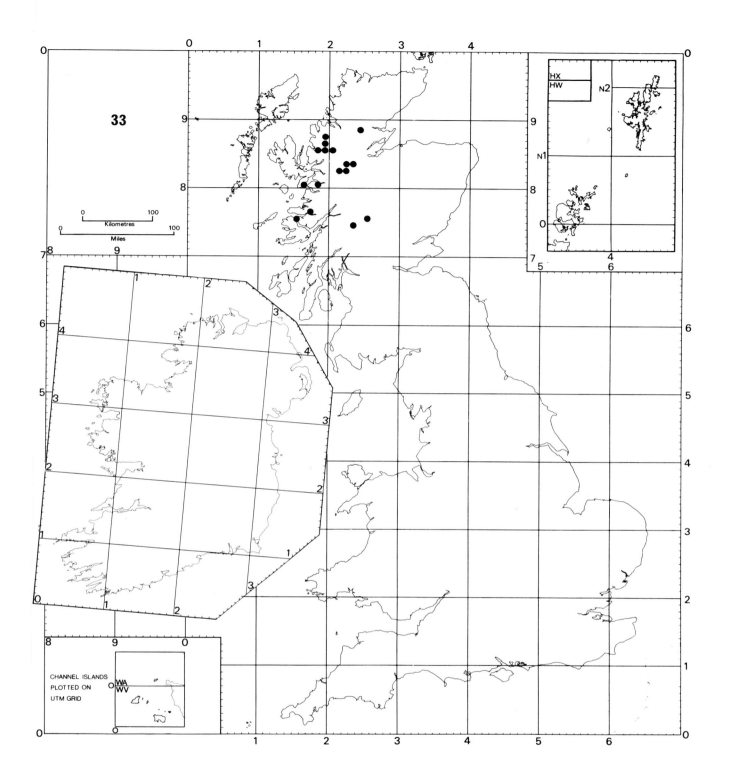

33 *Cavernularia hultenii* Degel.

A corticolous species, almost confined (except for localities in Skye and Mull) to relict areas of the Caledonian pine forest in the west and central Scottish Highlands; it occurs frequently on *Pinus sylvestris* ssp *scotica* in these relict forests, but is much more frequent there on *Betula* and *Sorbus aucuparia,* and also occurs on *Ilex.* It is, however, never abundant. In Skye, Mull, and near Loch Sunart it is on *Betula* in open woodlands. It occupies a niche in the *Pseudevernietum,* or sometimes in the west in the *Parmelietum laevigatae,* always in well-lit situations on arid, well-drained bark, with pH probably always below 4.5. It can be regarded as a relict of the Boreal period of pine-birch dominance, and though there is no direct evidence, its occurrence only in relict forest sites suggests that it was once far more frequent. It has not been found yet in the old pine woods in Scotland east of the Great Glen, and it appears to be limited to the more humid central and western areas of the Highlands. Not seen fertile in Britain. It is known widely in old pine and spruce forests in Norway and west Sweden but not further south or east and nowhere in planted pinewoods; it is a boreal-oceanic species.

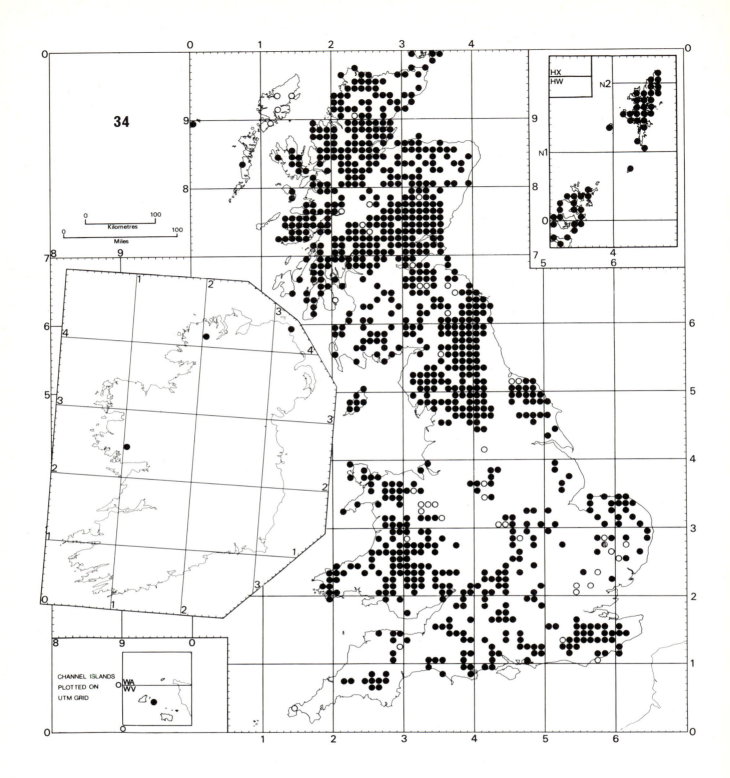

34 *Cetraria chlorophylla* (Willd.) Vainio

A member of the calcifuge *Pseudevernietum furfuraceae* occurring on substrates of pH 3.0-4.0 in most parts of England, Wales and Scotland, but very rare in Ireland; on twigs, branches and worked timber (especially fence rails), also on siliceous rocks in both this community and the *Parmelietum omphalodis*. Particularly characteristic of *Betula* twigs in the Scottish Highlands where it occurs with *C. sepincola* (map 37) and *Parmelia septentrionalis* (map 107). Widespread throughout the British Isles and absent only where sulphur dioxide levels exceed about 60 μg m^{-3}. Probably more widespread than in the nineteenth century due to increased bark acidification of deciduous trees (arising from increased sulphur dioxide levels). Separated from poorly developed specimens of *Platismatia glauca* in being greenish-brown (not grey) when wet. Primarily a boreal coniferous forest species in the northern hemisphere, with a western Eurasian-western North American distribution pattern.

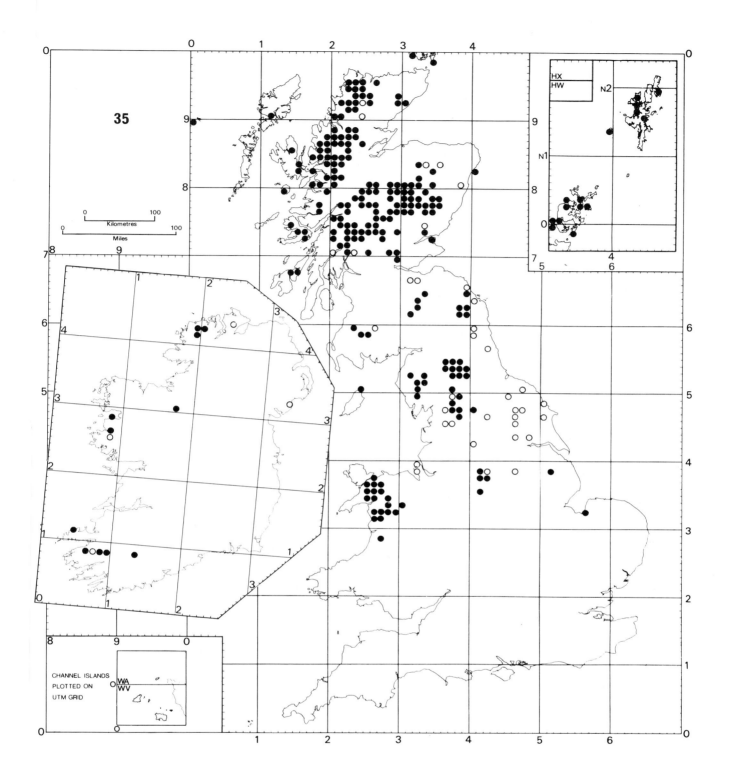

35 *Cetraria islandica* (L.) Ach.

An important and locally dominant component of upland lichen-rich heaths, including the *Arctoeto-Callunetum* and *Cladineto-Callunetum* growing amongst mosses, ericaceous shrubs, *Salix herbacea* or directly on peaty soil. Commonest in the Scottish Highlands but also recorded from lowland heaths in the Peak District, East Yorkshire, Lincolnshire and Norfolk. Evidently moderately tolerant of air pollution, but eliminated by the regular burning of grouse-moors. Circumpolar in the northern hemisphere, extending south to the southern Alps. The map includes records of the strictly upland subsp. *crispiformis* (Räsänen) Kärnef., with canaliculate to almost tubular lobes and only rare and indistinct laminal pseudocyphellae. There is also a distinct subspecies in Australasia, the southern tip of South America and the subantarctic, and another in Japan and other parts of eastern Asia. The closely allied *C. ericetorum*, confirmed only from Cairngorm in Britain but perhaps more widespread, differs in the consistently narrow lobes, medulla always being PD—, presence of different fatty acids, indistinct pseudocyphellae, and rich apical branching of the lobes.

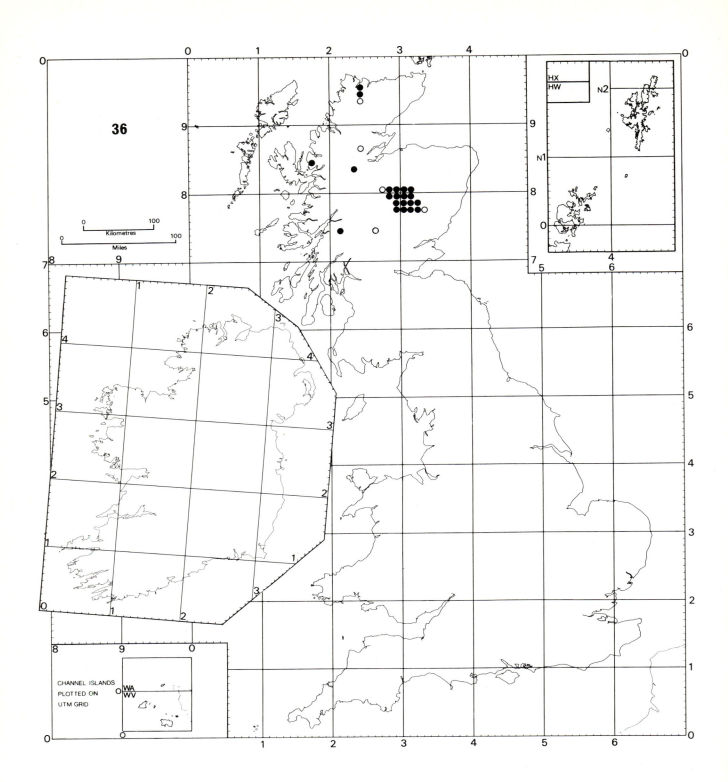

36 *Cetraria nivalis* (L.) Ach.

One of the characteristic and most conspicuous species of lichen-moss heaths in the Scottish Highlands, occurring in *Calluna-Vaccinium uliginosum-Empetrum-Salix herbacea* communities, avoiding patches with late snow-lie, and forming a distinct altitudinal band at about 900-1050 + m on Cairngorm. Only likely to be adversely affected by habitat destruction due to excessive skiing in the foreseeable future. Completely circumpolar in the northern hemisphere, mainly in the arctic zone extending north to about Lat. 83°N, but extending southwards in Europe to the Alps.

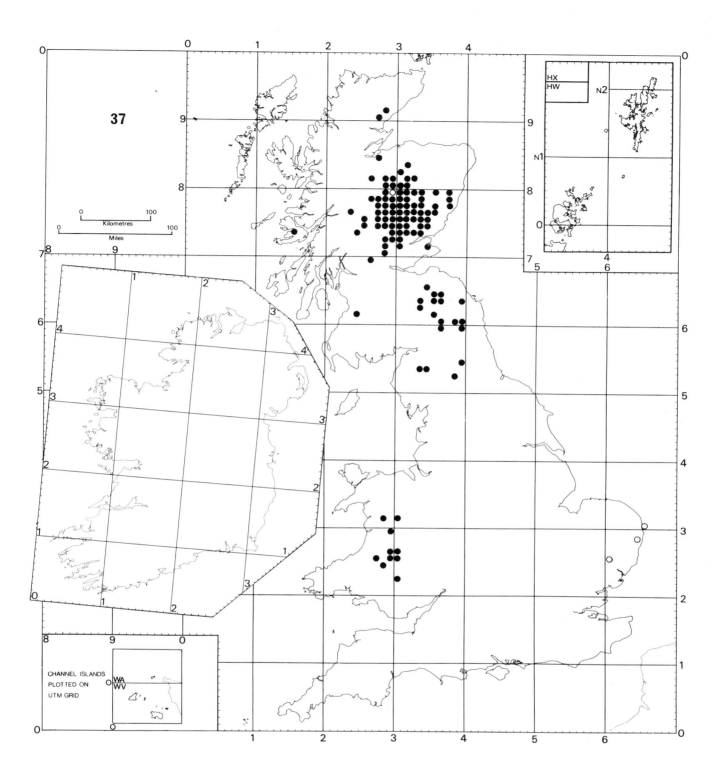

37 *Cetraria sepincola* (Ehrh.) Ach.

A northern and somewhat eastern species locally abundant in the Scottish Highlands but also known from northern England and central Wales where it is rare. It occurs characteristically on well-lit *Betula* twigs together with *Cetraria chlorophylla* and *Parmelia septentrionalis* in a variant of the boreal-continental *Parmeliopsidetum ambiguae,* but has also been discovered on fence posts in exposed situations. Usually fertile, which aids its separation from *C. chlorophylla,* a species that also differs in the much paler brown colour when dry. Probably circumboreal in the northern hemisphere, with somewhat continental tendencies, especially common in the boreal coniferous forests of Scandinavia and North America; also present in upland areas such as the Ardennes, as well as in similar situations from the Alps to the Balkans.

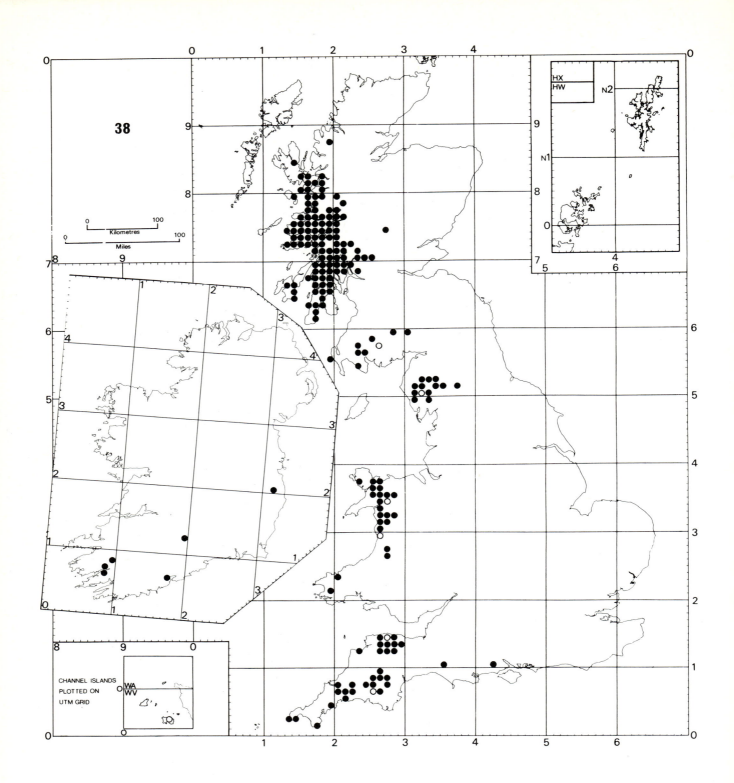

38   *Cetrelia olivetorum* (Nyl.) Culb. & C. Culb.

A distinctly western species in the British Isles found in constantly humid, mild and moderately well-lit sheltered broad-leaved woodlands, especially in valleys, often amongst mosses. Primarily a species of the calcifuge *Parmelietum laevigatae,* but also an important component of the euoceanic facies of the *Lobarion pulmonariae;* in similar communities on sheltered mossy rocks, usually within woodlands. Several chemotypes are known but are best interpreted as races of a single southern boreal to temperate species rather widely distributed in western, central and eastern Europe extending northwards into central Norway and in southern Europe forming an association with *Pseudevernia furfuracea* in *Fagus-Abies* forests, especially above 1400 metres. Also known from temperate areas of south-east Asia and eastern North America.

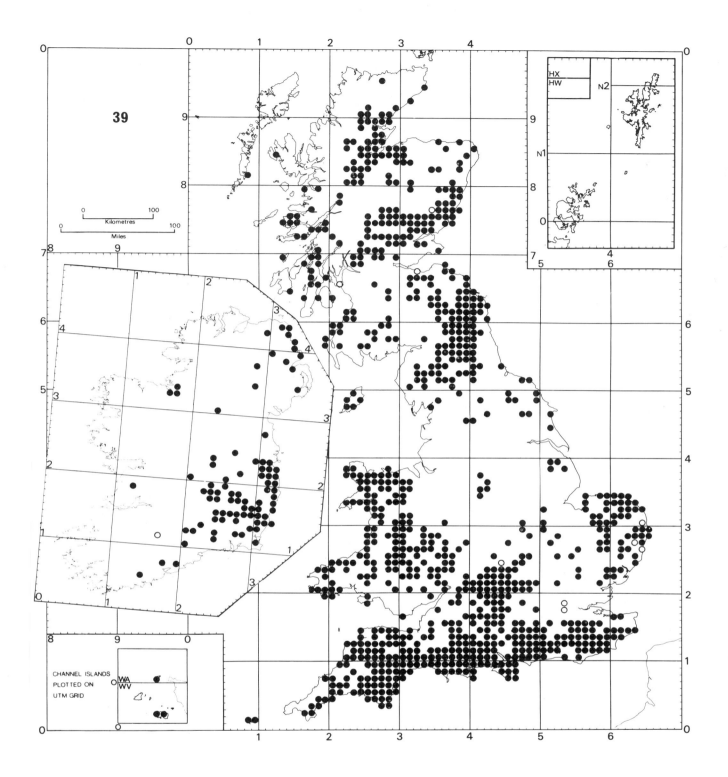

39 *Chrysothrix candelaris* (L.) Laundon

A widespread, locally common, crevice species characteristic of the *Calicietum hyperelli,* a species-poor association on dry, rough or moderately smooth bark of hardwood trees in lowland areas and also on coniferous trees in central and eastern Scotland; occasionally found on dry sheltered siliceous rock faces. Widespread in many areas of Britain especially in north-east and southern England and tolerant of up to 60 μg m^{-3} sulphur dioxide. Absent from the Midlands and central lowlands of Scotland due to air pollution and farm sprays; also becoming scarce in the west of Scotland and Ireland probably due to too moist a climate; in the driest parts of Scotland it may dominate the south sides of the trunks of woodland oak. Worldwide in distribution.

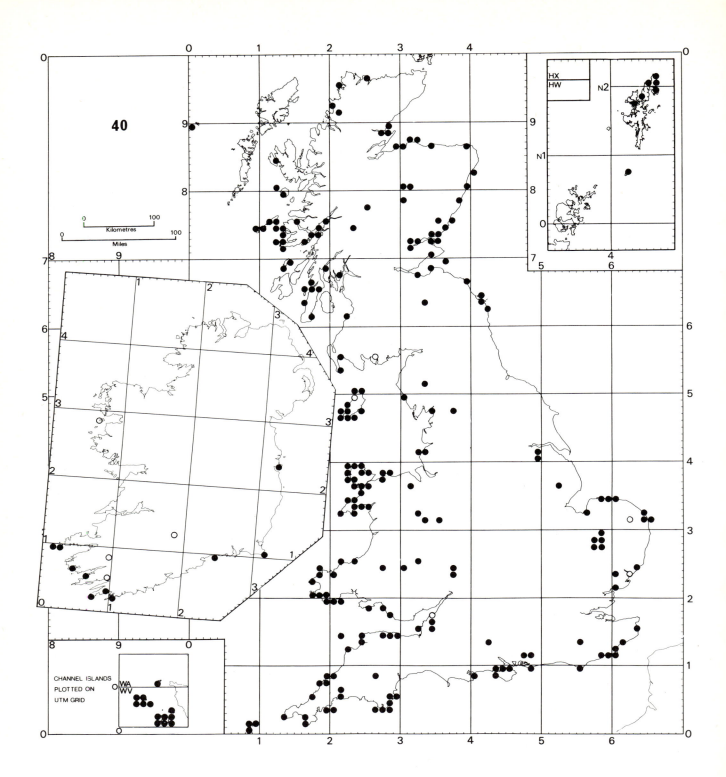

40   *Cladonia foliacea* (Huds.) Willd.

A characteristic species of short moss-lichen swards developed on well-drained calcareous coastal soils and humus stabilized sand dunes and shingle, in wind-swept sunny situations, sometimes in crevices of both siliceous and calcareous rocks of bird roosting cliffs. In suitable sites around all the British coasts in the xeric supralittoral and terrestrial zones, tolerating a high degree of salinity, and rarely also on inland peaks affected by salt spray or in highly calcareous sites, (e.g. Breckland). A component of the *Cladonietum alcicornis* subject in both coastal and inland sites to damage by trampling and, in less exposed situations, overgrowth by grasses when any grazing pressure is reduced. Not uncommonly fertile, at least in south-west England, but the podetia are dwarfed by the squamules and so easily overlooked. Formerly often confused with *Cladonia cervicornis* but readily distinguished by the yellow-green lobes. Mainly confined to Europe, with oceanic tendencies, extending from the Mediterranean to south-west Finland; absent from North America.

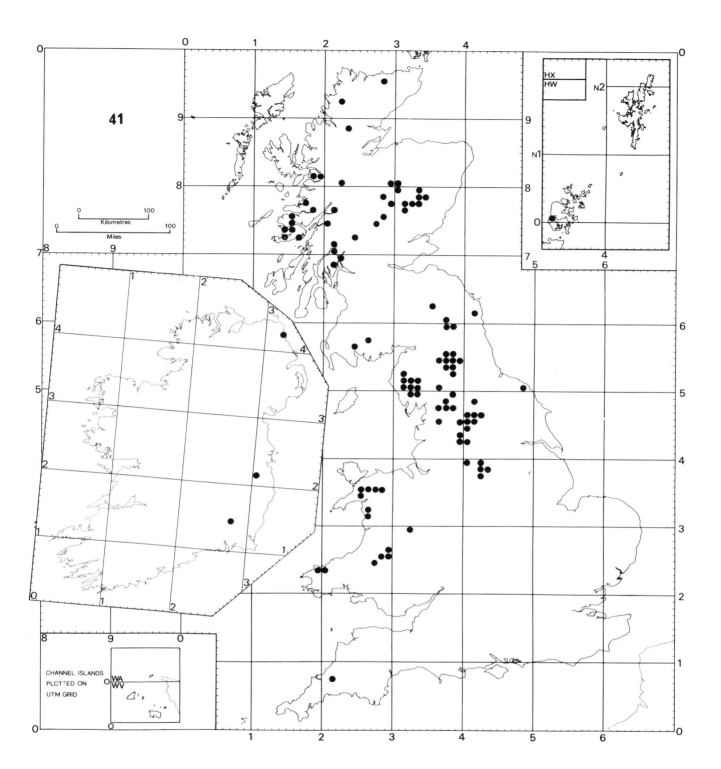

41   *Cladonia luteoalba* A. Wilson & Wheldon

A very local species, usually occurring as only a few small clumps, on highly oligotrophic peaty soils or humus on rock ledges in exposed upland heaths; also rarely corticolous at the bases of old trees. Most frequent in the British Isles in the Lake District and Pennines but extending south through Wales to Bodmin (Cornwall). The species could be affected by the burning of moorland or changes in grazing regimes. It is easily recognised by the large squamules, yellowish and cottony arachnoid below. Although first described from the British Isles in 1909, and evidently more frequent then, this species proves to have a wide distribution in suboceanic areas of north-west Europe, and is also known from Alaska, the Austrian Alps and Spitsbergen; perhaps formerly circumpolar in the nothern hemisphere but eliminated from many areas as a result of the last glaciations. There are two chemical races in Britain.

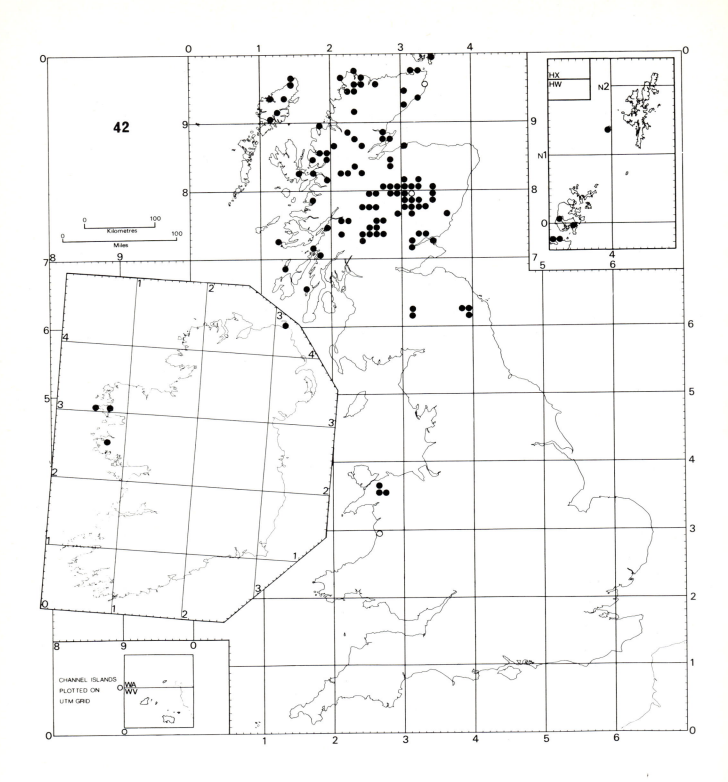

42 *Cladonia rangiferina* (L.) Wigg.

A species of exposed moss-lichen heaths most abundant in the Scottish Highlands where it can be an important component of the vegetation (e.g. on Cairngorm). Also occurring near sea level often on cliff ledges in suboceanic areas in Shetland, Orkney, Caithness and western Scotland. Distinguished from allied species by the bluish-grey to pale grey colour, abundant branching and K+ yellow, PD + red reactions. The species probably benefits from grazing and it is eaten by sheep as well as reindeer. Little is known of the former range of this species in the British Isles; literature reports from lowland Britain are errors based on allied species. *Cladonia rangiferina* s.str. is a common and wide-ranging circumboreal species with outlying localities in the Himalayan region and Japan. Descending onto lowland heaths in Denmark, central France and the Netherlands.

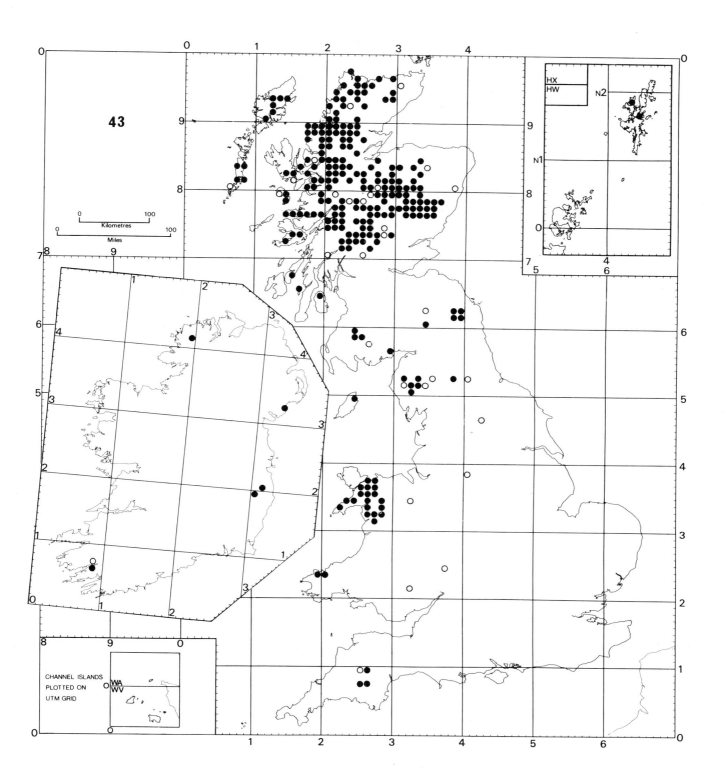

43   *Cornicularia normoerica* (Gunn.) Du Rietz

An arctic-alpine species restricted to well-lit coarse-grained siliceous rocks (especially granite) in wind-swept situations. Occurring at lower altitudes in northern Scotland and Shetland but largely a mountain-top species in the north of England and Wales extending southwards to Dartmoor where it is now very rare. It is confined to macrolichen-dominated communities of the *Umbilicarietum cylindricae* and may be part of a seral stage progressing to the *Rhizocarpetum alpicolae*. Widespread in northern and mountainous areas of Europe, but often rather local with oceanic tendencies, extending south to Portugal and Macedonia; rare in North America where it is largely restricted to the west coast; also known from Japan.

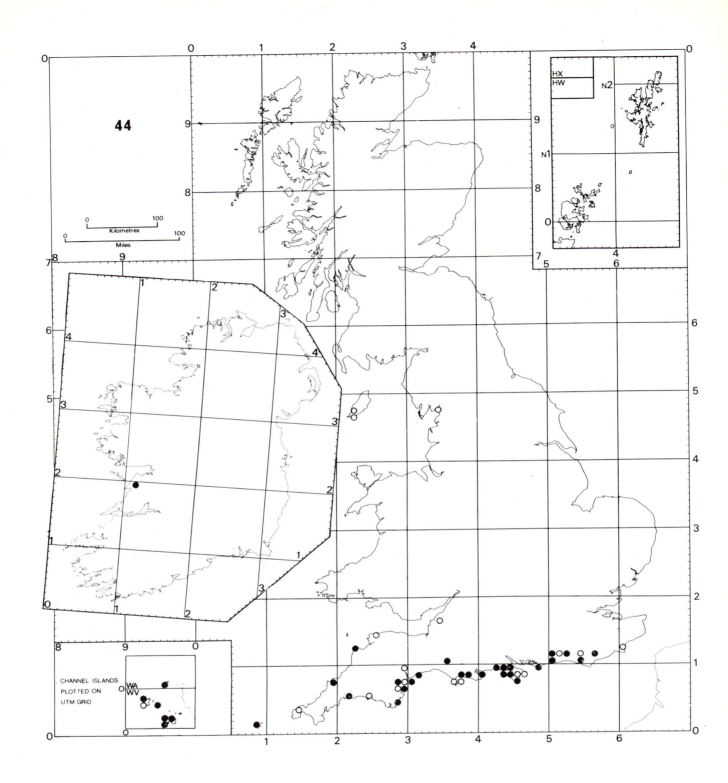

44 *Cryptolechia carneolutea* (Turner) Massal.

A rare species of shaded boles of old trees with a nutrient-rich bark, especially *Ulmus,* more rarely *Acer* and *Fraxinus,* in parkland and wayside habitats. Often covering extensive areas and the dominant species of the *Gyalectinetum carneoluteae* of the *Xanthorion parietinae* alliance. Now confined to a few sheltered, warm situations in extreme southern England, normally near the coast; previously recorded as far north as the Isle of Man and north Lancashire (Silverdale). A further recent dramatic decrease has occurred due to Dutch Elm Disease and pollution by inorganic fertilizers. Southern-oceanic in Europe, in similar habitats.

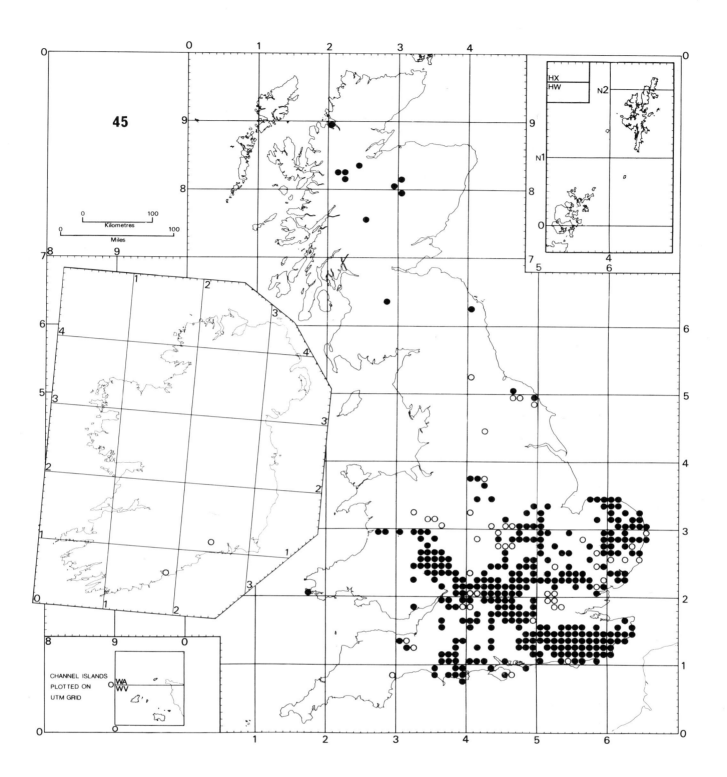

45 *Cyphelium inquinans* (Sm.) Trevisan

This species has a broadly south-eastern distribution in the British Isles, mainly occurring on old fence posts and rails, gates and other sawn timber and more rarely on old dry bark in well-lit open situations. It commonly occurs in association with *Lecanora conizaeoides* Nyl. ex Crombie and *L. varia* (Hoffm.) Ach. forming a distinctive community, the *Cyphelietum inquinantis*. In the Scottish Highlands it is restricted to ancient pine forests, occurring there on both bark and decorticate wood — natural habitats for the species over most of its range. Moderately tolerant of sulphur dioxide pollution but absent where mean values exceed about 70 μg m^{-3}. A decline in the number of wooden fences and spraying of agricultural chemicals may account for its scarcity in parts of East Anglia and Lincolnshire in which it would otherwise be expected. A circumboreal species with continental tendencies, particularly common in Scandinavia and also present in the mountainous and northern lowland areas of Europe.

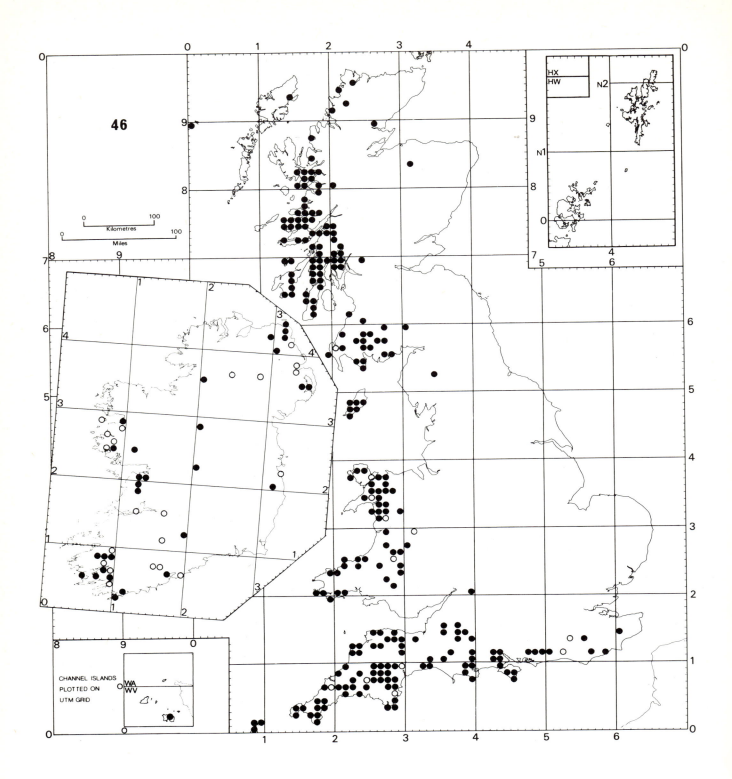

46 *Dimerella lutea* (Dickson) Trevisan

A distinctly southern and western species found overgrowing mosses on bark, on siliceous rocks or encrusting decaying grass tussocks and humus-rich soils in consistently humid and somewhat shaded situations, avoiding areas with marked temperature fluctuations; sometimes more or less directly on bark. Almost restricted to the *Lobarion pulmonariae* (western facies) and pre-*Lobarion* pioneer communities and an indicator of sites with a long history of ecological continuity. Perhaps formerly more widespread in lowland sites but poorly recorded in the past. Absent from areas where mean sulphur dioxide levels exceed about 30 μg m^{-3} and also subject to loss through woodland disturbance. Essentially pantropical but extending into oceanic and suboceanic temperate regions in the northern hemisphere, north to western France; reported from southern Scandinavia; also present in eastern North America.

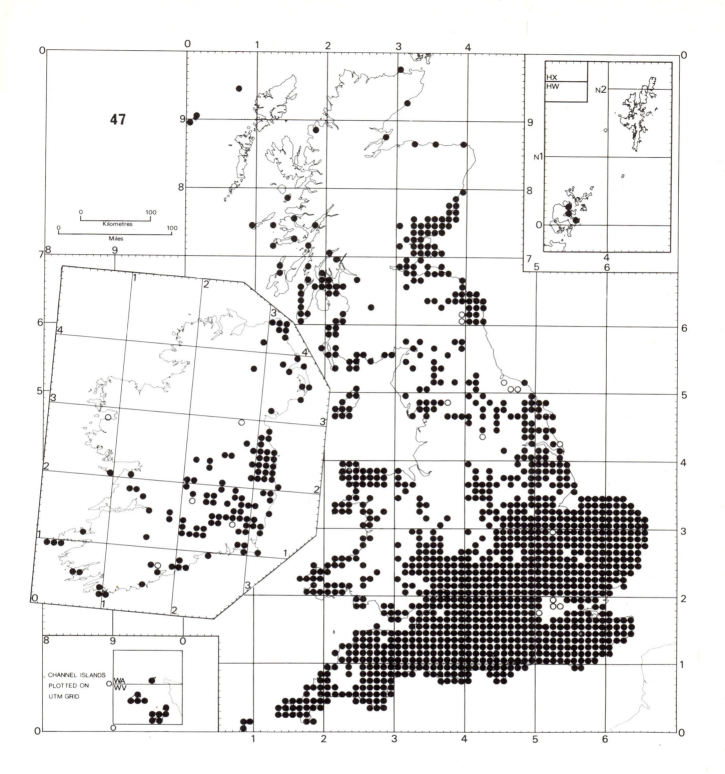

47 *Diploicia canescens* (Dickson) Massal.

D. canescens occurs on both rocks and trees; it is characteristic of dry, often sheltered, shaded nutrient-enriched habitats in coastal and inland lowland sites. Phytosociologically *D. canescens* is an important associate in several diverse communities. When corticolous it enters the *Arthonietum impolitae* on dry sides of trees in pastures; the *Lecanactidetum premneae* at the bases of ancient *Quercus* in parkland; the *Buellietum punctiformis* on nutrient-enriched or hypertrophicated bark (e.g. *Ulmus, Acer pseudoplatanus, Fraxinus*) by dusty roadsides and in hedgerows, especially in areas of moderate sulphur dioxide pollution up to 60 μg m^{-3}; and the *Physcietum ascendentis* on wayside trees and its saxicolous equivalent on basic rocks. When saxicolous it enters the *Caloplacetum heppianae* on limestones and man-made substrates; the *Candelarielletum corallizae* on ornithocoprophilous rocks; and the *Ramalinetum scopularis* on sheltered coastal rocks. It is more frequent in south-east England south of a line drawn between the Severn Estuary and the Wash, especially in churchyards; further north it becomes rare and more restricted to sunny, warm, coastal sites and dry crevices in sheltered tree boles, bases of old mortar walls, and limestone outcrops, extending to Orkney in such habitats. Widely distributed in Atlantic and Mediterranean parts of Europe but rarer in continental areas. Also recorded at least from north Africa, North America and New Zealand.

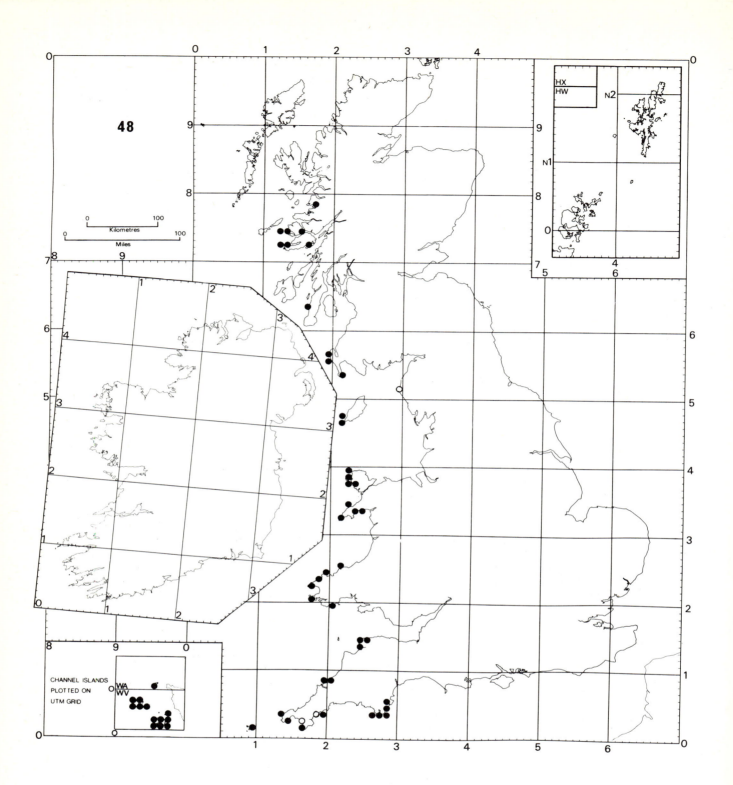

48 *Diploschistes caesioplumbeus* (Nyl.) Vainio

A local, exclusively maritime species on sunny, hard, siliceous rocks on moderately sheltered but sunny shores. Xeric-supralittoral, it belongs to a facies of the *Ramalinetum scopularis* dominated by crustose lichens which resembles the inland *Lecanoretum sordidae* with additional, exclusively maritime species. It occurs most commonly in lichen mosaics on small, low rocks or stabilised boulders on thin soil just above the h.w.m.s.t. level on the shore and particularly favours a south-facing aspect; it does not normally colonize larger outcrops and cliffs above the sea. In south Devon it has been found parasitising *Lecanora gangaleoides,* taking over its algal partner. A markedly western species in Britain sometimes locally common on islands and peninsulas from south Devon to Argyllshire. Abroad it is known from France and the Iberian Peninsula; its distribution further afield is unreliable due to confusion with closely related species. The genus is poorly understood and in urgent need of a monographic treatment.

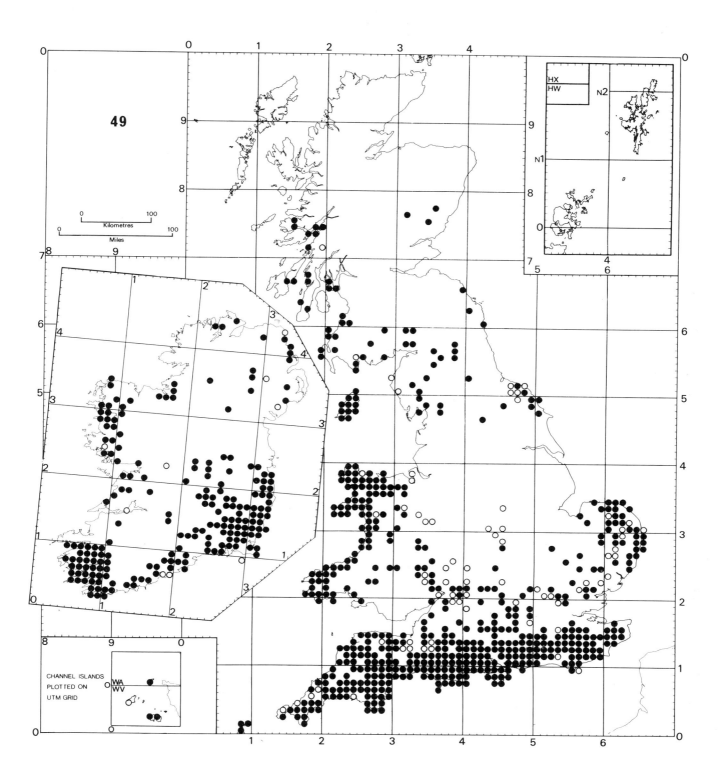

49 *Enterographa crassa* (DC.) Fee

A general southern species in the British Isles becoming scarce in northern England and Scotland though extending as far north as Appin and the Island of Mull. Its absence from large areas of central England must be attributed to a combination of air pollution and the clearance of old woodlands. Its mosaics occur, often in extensive pure stands, in communities of the *Pyrenuletum nitidae* on a wide range of smooth-barked deciduous trees (pH 5.1-5.8) growing in deep shade, also on rough bark in the south and very rarely on shaded siliceous rocks. Through most of its range it acts as an indicator of sites with a long history of ecological continuity but this correlation is less well marked in the extreme south-west, where its powers of dispersal are evidently optimal. Very variable in colour and the bullateness (thickness) of the thallus; the similar extinct *Enterographa elaborata* differs in the large spores and PD+ yellow-orange reaction. Southern-oceanic in Europe, extending north to southern Sweden, also present in western Portugal, Yugoslavia, and south to the Azores; reports from other areas are dubious.

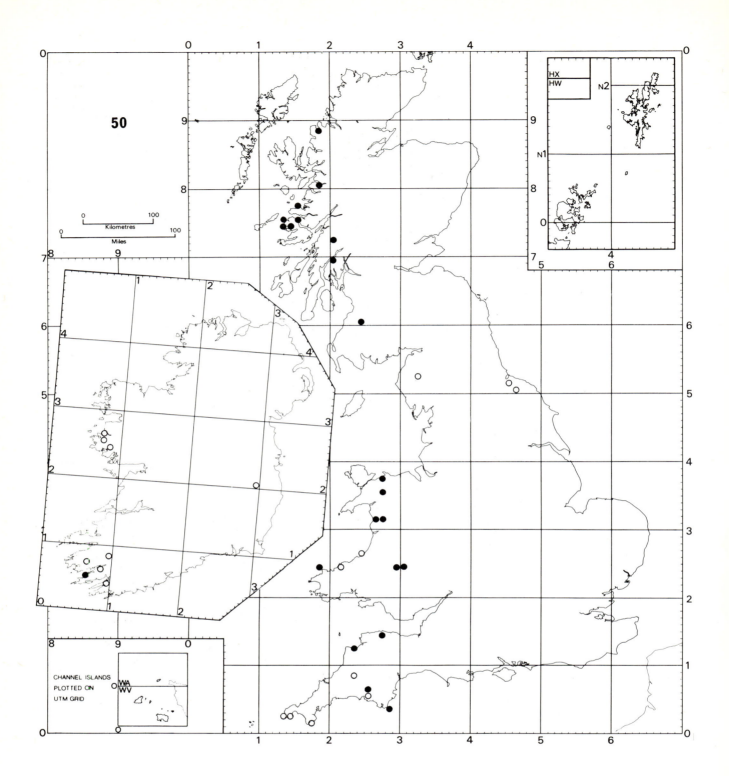

50 *Enterographa hutchinsiae* (Leighton) Massal.

An extremely local, probably under-recorded and mainly western species found primarily on continuously shaded underhangs and recesses on more or less smooth grained siliceous rocks from Devon to W. Inverness. Faithful to the *Opegraphetum horisticogyrocarpae,* often within woodlands, but also associated with seepage tracks on coastal rocks and occasionally almost aquatic by the edges of streams and rivers; exceptionally corticolous on exposed tree roots and buttresses by shaded, sheltered stream sides. It has probably changed little in its overall distribution but could easily be lost from many of its sites through disturbance and opening up of woodlands in which it occurs. Rare saxicolous specimens of *Enterographa crassa* differ in the punctiform ascocarps and pale greenish to reddish brown thalli. First described from south-west Ireland but subsequently also reported from western Germany, northern Austria and Rumania; apparently confined to oceanic and suboceanic areas of Europe.

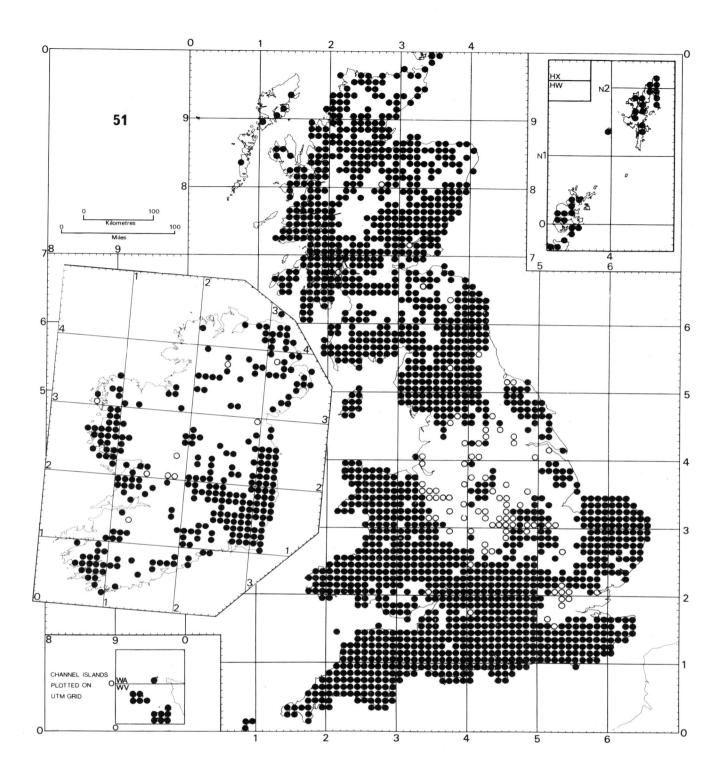

51 *Evernia prunastri* (L.) Ach.

Widespread throughout the British Isles, on well-lit bark of deciduous trees, mainly in the pH range 4.5-6.0, but also present on ericaceous shrubs and fence posts (the main habitat in northern Scotland and Shetland), exceptionally on nutrient rich rocks, stabilized shingle, brick or short turf (in coastal situations and dunes). Characteristically occurring in the *Parmelietum revolutae,* but also in a wide range of other communities including the *Lobarion pulmonariae* and *Ramalinetum fastigiatae.* The species has recently returned to the London area, but has declined in large areas of the Midlands where mean sulphur dioxide levels exceed about 70 μg m^{-3}. Formerly regularly fertile, but apothecia now produced only where the mean sulphur dioxide levels are less than about 30 μg m^{-3}. A grey usnic-deficient chemical race may co-exist with the normal yellow-grey type. Favoured nest-building material of the Long-tailed Tit *(Aegithalos caudatus).* A boreal-temperate species widespread throughout Europe; also common in western North America and extending to north Africa and Japan, but absent from the southern hemisphere.

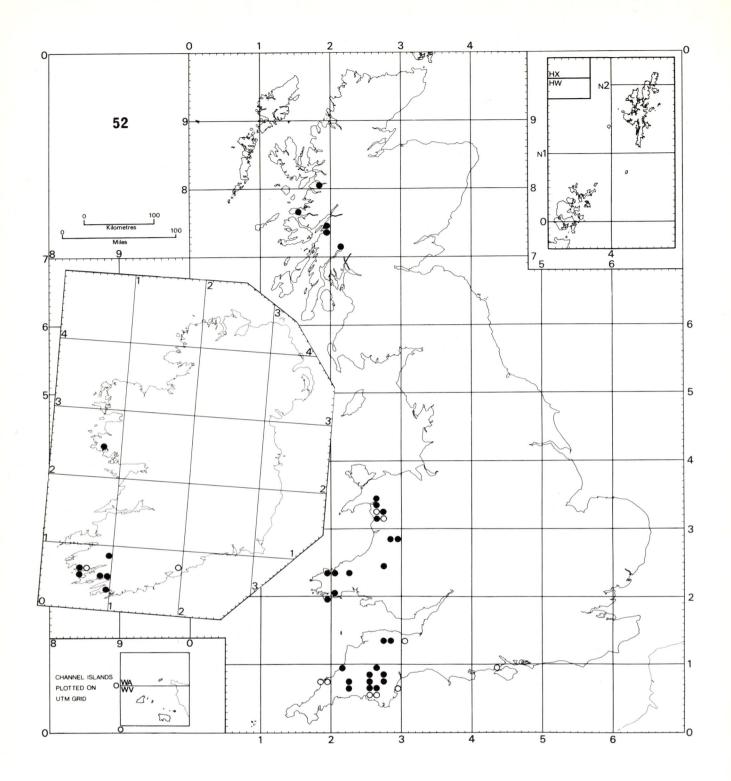

52 *Graphina ruiziana* (Fée) Müll. Arg.

A predominantly south-western species, commonest in the high-rainfall woodlands on Dartmoor, but extending north as far as west Inverness in sheltered sites, and formerly east to the New Forest. The species requires both a high rainfall and humidity and a mild climate (within the 5°C January and 15°C July mean isotherms). It occurs on smooth bark with a low pH on a wide range of trees, but is perhaps commonest on *Ilex aquifolium* and *Sorbus aucuparia* in stands allied to the *Graphidion scriptae.* On moderately shaded and sheltered to well-lit exposed situations where the humidity permits this; in the latter case often on younger branches of trees with the *Parmelietum laevigatae* on their trunks. In Europe otherwise only known from Brittany, but there are scattered literature reports suggesting a pantropical distribution, in which case the British Isles is the northernmost extension of its range.

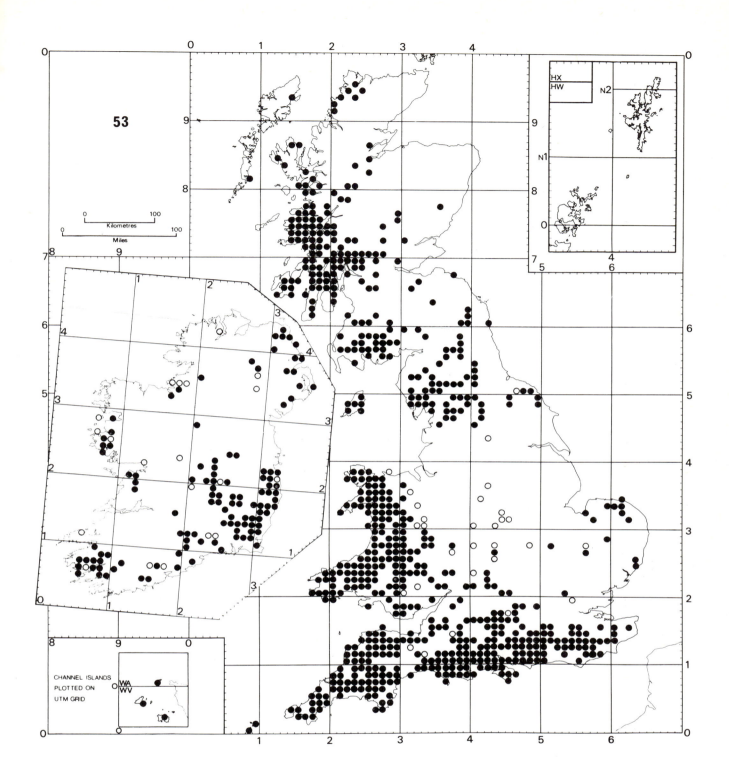

53 *Graphis elegans* (Sm.) Ach.

A characteristic species of the *Graphidion scriptae,* occurring on moderately shaded smooth bark, especially branches, twigs and young trees; the phorophyte range is wide, and it can occur on *Calluna* in coastal situations. Widespread in western and southern parts of the British Isles, but becoming very rare to absent in eastern Scotland. It has probably been lost from most of central England due to a combination of air pollution and modern farming methods. A distinctly oceanic species in Europe, extending northwards into north-west Germany and the extreme south of Scandinavia (where it is very rare). Published records suggest that the species is cosmopolitan, but many are probably erroneous; reliably reported, however, from both the Dominican Republic and New Zealand.

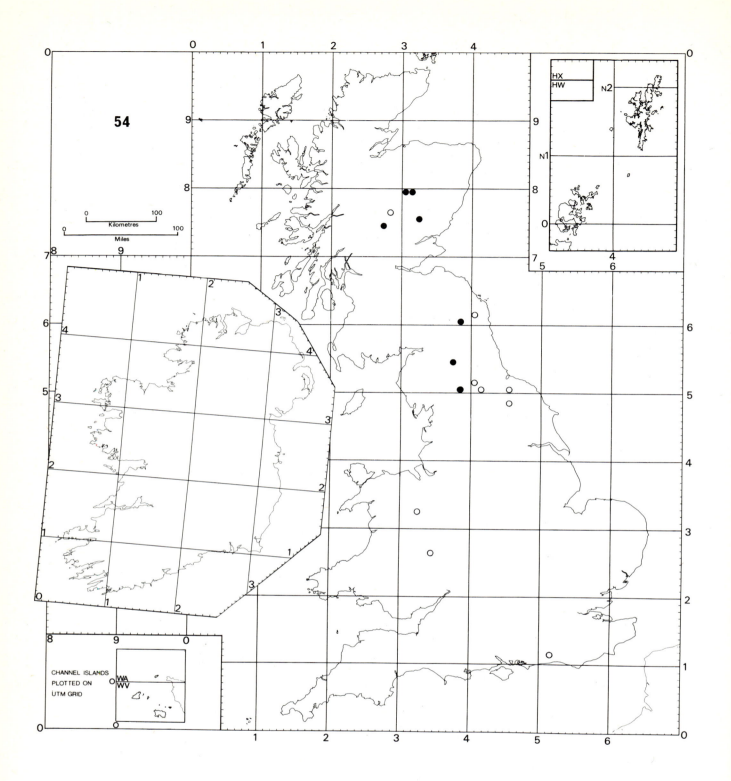

54 *Gyalecta ulmi* (Swartz) Zahlbr.

A rare and local species over-growing soil and mosses on sheltered, more or less basic rocks, and on boles and bases of old *Ulmus*. It is recognised by the conspicuously white and regularly crenate margins of the apothecia. It has disappeared from most English localities due to habitat destruction and Dutch Elm Disease. Only known from a few sites in northern England and the central highlands of Scotland, avoiding high rainfall areas of the west. Widely distributed, but scattered over much of Europe, including Iceland and Caucasia.

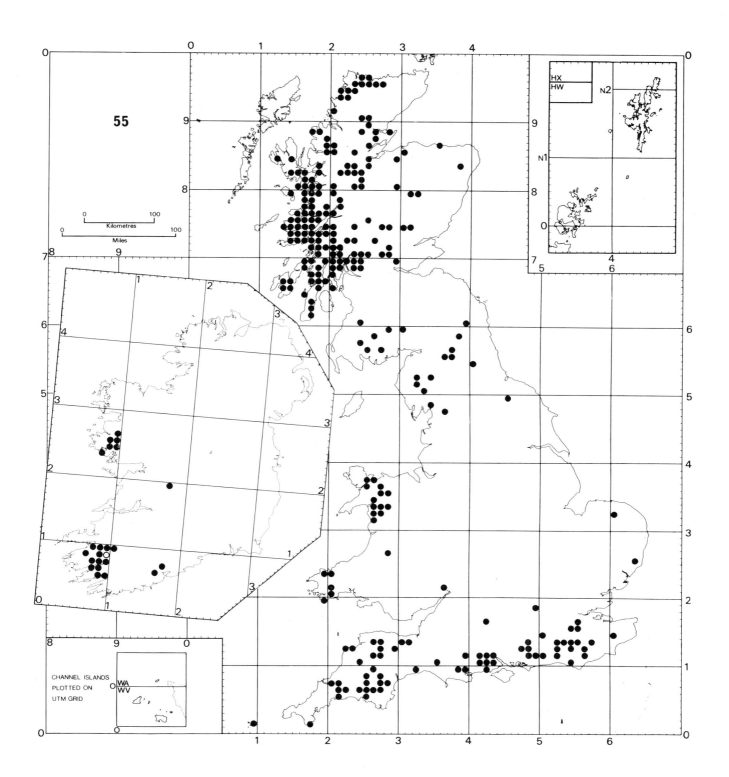

55 *Haematomma elatinum* (Ach.) Massal.

A species of acid or leached tree boles, most frequently *Betula, Salix, Fagus, Ilex, Sorbus aucuparia* and *Quercus*. In southern England it is very local and confined to sheltered, often boggy sites in old woodlands; in western areas it is less restricted and sometimes locally abundant and is often found in sheltered carr and sheltered moist ravines. It belongs to the *Graphidion scriptae,* where it merges into the *Pertusarietum amarae,* and a moist facies of the *Lecanoretum subfuscae.* Widely distributed in the north temperate zone in Europe and North America.

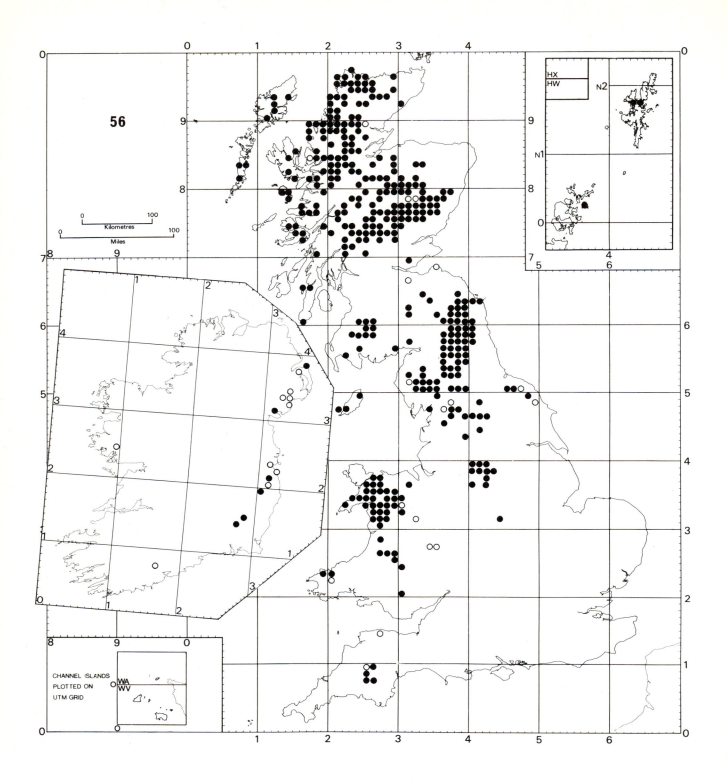

56 *Haematomma ventosum* (L.) Massal.

An upland species restricted to well-lit, especially coarse-grained hard siliceous rocks in exposed situations. This species only exceptionally colonises man-made structures and is intolerant of nutrient-enrichment (e.g. bird droppings). Thalli may be either bright yellowish-green or greyish (± usnic acid), and both chemotypes regularly grow together; it is normally fertile. Most frequent in the Scottish Highlands, northern England and Snowdonia, but extending south through the Pennines to the Peak District and Charnwood Forest, with a few localities in the south-west on Dartmoor. It enters the arctic-alpine *Rhizocarpetum alpicolae* in Cairngorm, but in most of its British localities it is an important component of the *Parmelietum omphalodis*. Its absence from lowland Britain is due to the lack of suitable natural rock outcrops. Apparently endemic in Europe, where it occurs through most of Scandinavia, the Alps, Carpathians and Pyrenees. The more pronounced arctic-alpine *H. lapponicum* Räsänen (Pd—), apparently extinct in Britain, is circumpolar in the northern hemisphere.

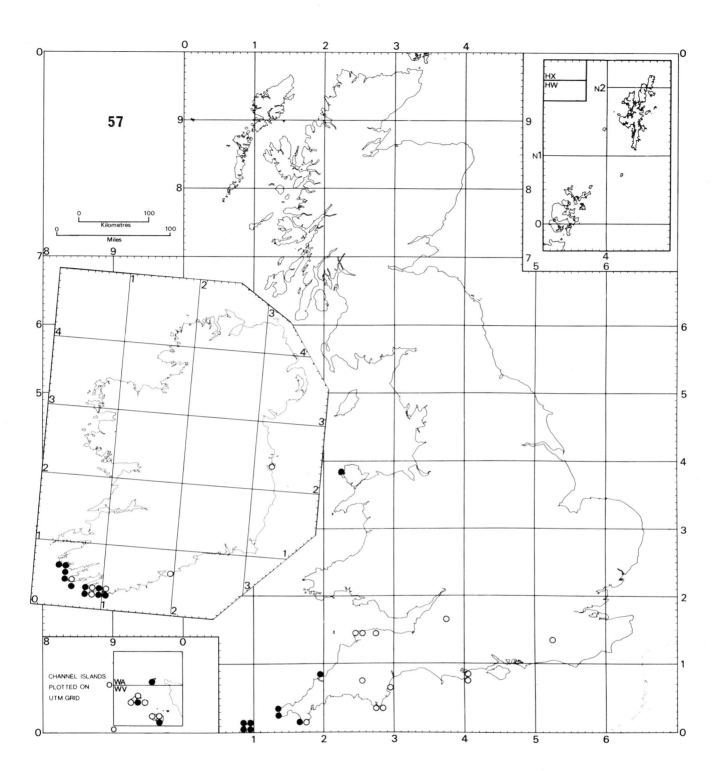

57 *Heterodermia leucomelos* (L.) Poelt

An extremely rare and declining species of sunny exposed, mainly coastal cliff-tops, moss-lichen turf or mossy rocks, now confined to the Channel Islands, Isles of Scilly and Cornwall, with an outlying station on Anglesey. Formerly also present in similar situations in Devon, extending eastwards into Sussex, but affected by over-collection, burning and severe trampling. A widespread and pantropical species of the New and Old Worlds. Oceanic and western in Europe with the British Isles representing the northernmost extension of its range. In western France (Brittany) it is corticolous on boughs and twigs in the crowns of trees in association with *Teloschistes flavicans;* it was formerly corticolous in England.

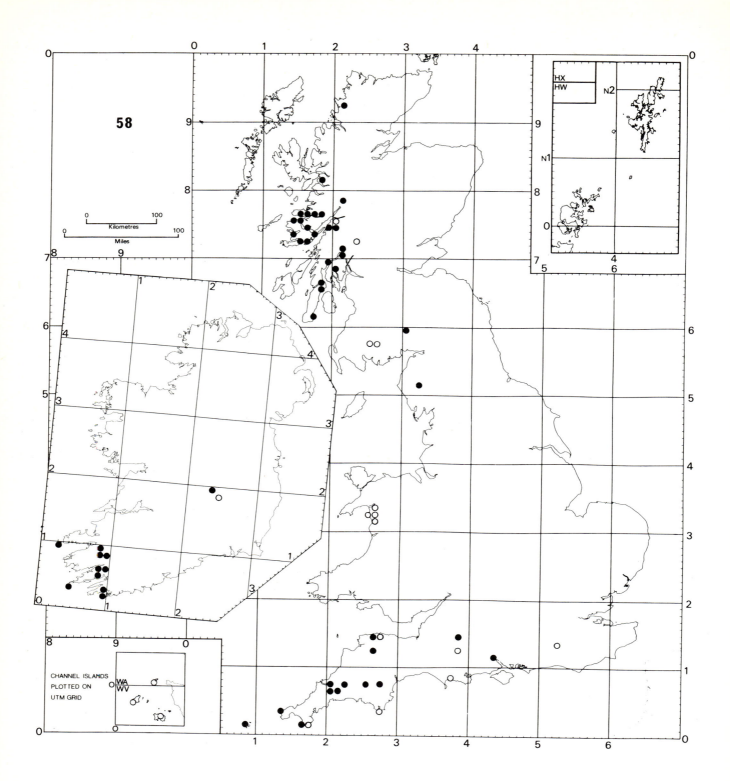

58 *Heterodermia obscurata* (Nyl.) Trevisan

A southern and eastern species, extremely rare in the south of England and central Ireland but becoming locally abundant in some of its western Scottish sites, and still fertile in the Cowal Peninsula, Argyll. A characteristic species of ancient parkland or other areas with a long history of ecological continuity, usually occurring on mossy boles and high up on the main branches of trees in sheltered humid situations; also present in short turf and on mossy rocks in coastal sites affected by persistent sea-mist. In western Scotland (and Devon) it is often optimally developed in *Salix* carrs or on aged *Corylus* in sheltered valley sites. The species has been lost from some of its coastal sites in England, perhaps mainly due to over-collecting, and also trampling, the destruction of ancient woodlands and drainage of established carrs; air pollution also contributes to its current rarity. A pantropical species of both the New and Old Worlds; present in Asia, Australasia, Africa, and North and South America. The British localities represent the most northerly extent of its highly southern-oceanic range in Europe.

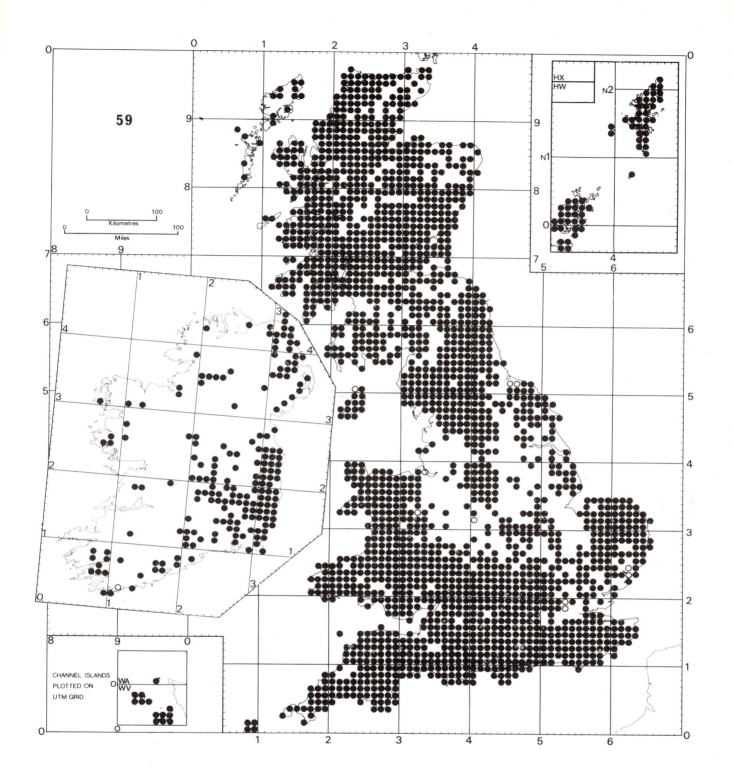

59 *Hypogymnia physodes* (L.) Nyl.

With *Parmelia sulcata* (map 111), this is one of the most ubiquitous macrolichens in the British Isles. It is absent only where mean sulphur dioxide levels exceed about 100 μg m⁻³, and responds rapidly to fluctuating levels of this pollutant. It occurs on a wide range of substrates, from supralittoral to montane, provided that they are moderately acid (probably with an optimum pH of about 4.5), on siliceous rocks as well as trees, and is especially common on *Calluna* stems and twigs of broad-leaved and coniferous trees. A member of numerous associations, but most important in the *Pseudevernietum furfuraceae*. A temperate to sub-arctic species, circumpolar in the northern hemisphere, and widespread throughout Europe, though becoming purely montane towards the south; known also from Kenya and New Zealand, but literature reports for other areas require confirmation.

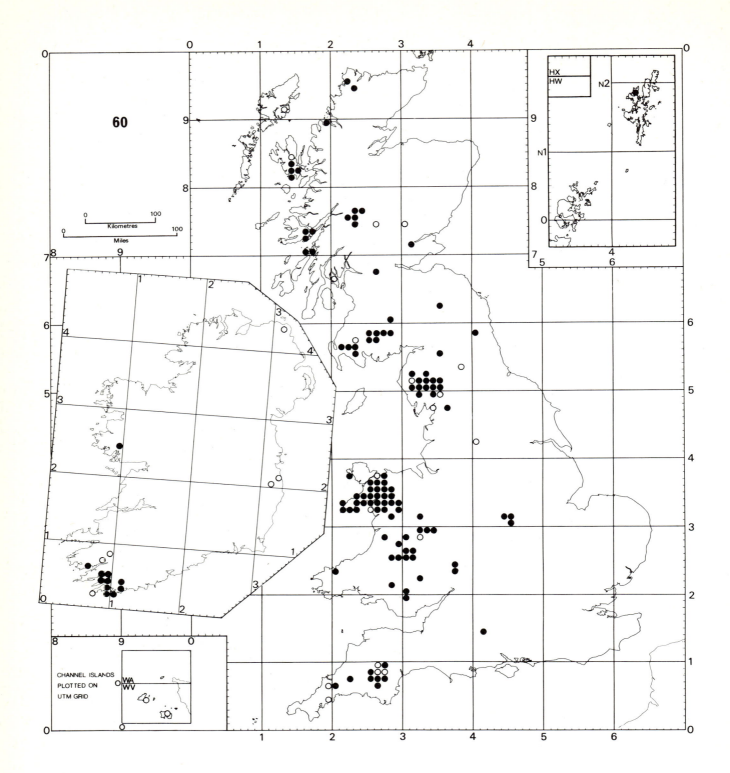

60 *Lasallia pustulata* (L.) Mérat

A distinctive species characteristic of nutrient-enriched, often mineral-rich, acid rocks and boulders. The species often occurs on erratics or outcrops with moderate hypertrophication from farm animals, particularly sheep, or bird droppings. Sometimes it forms extensive and spectacular 'swards' in nutrient-rich rain tracks down more or less steeply-inclined rock faces. It occurs in a nutrient-enriched facies of the *Fuscideetum kochianae* but is characteristic of the *Parmelietum glomelliferae.* It is a rather local species, often represented in an area only on a single boulder. It is most frequent in the Lake District, north Wales and Dartmoor, Devon. It is apparently moderately tolerant of air pollution but sensitive to inorganic fertilizers. Widely distributed throughout the north temperate zone, including Asia and North America, but also present in the Canary Islands; north to central Sweden, but avoiding extreme arctic regions.

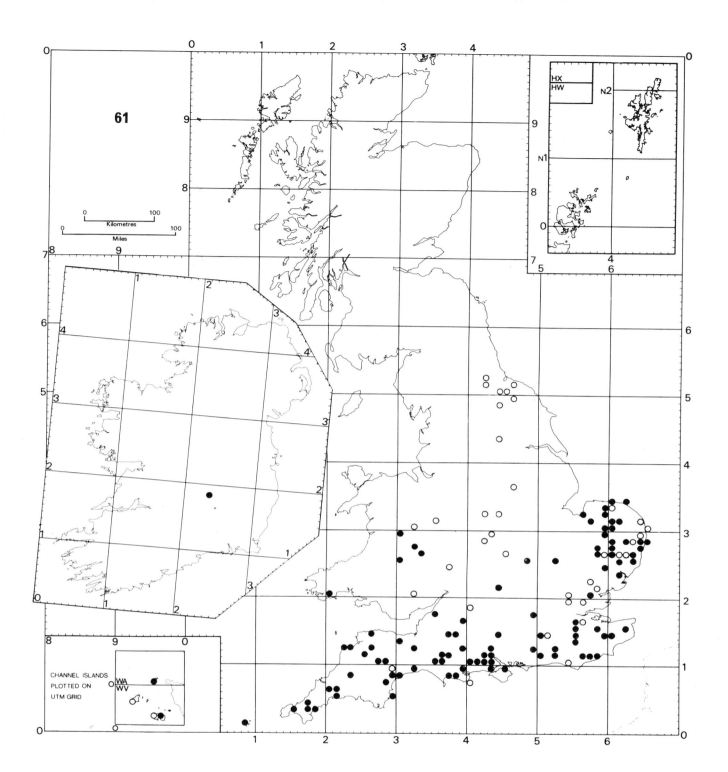

61   *Lecanactis lyncea* (Sm.) Fr.

One of the characteristic species of the *Lecanactidetum premneae*, a post-climax association of very ancient trees, mainly *Quercus*, occurring on dry and brittle bark not affected by run-off or direct rain. *L. lyncea* is typically found in sunny situations, and very commonly just below where major boughs fork from the trunk. An old woodland indicator species able to persist in open parklands, its status has been affected both by the felling of ancient trees and air pollution (absent where mean sulphur dioxide levels exceed about 60 μg m^{-3}). A local southern and eastern species in the British Isles, it is only known from western Europe and north Africa, extending north only to Denmark and southern Sweden where it is very rare. In western Europe generally it is far rarer than in southern England, due apparently to the general lack of very old oaks in most of the forests.

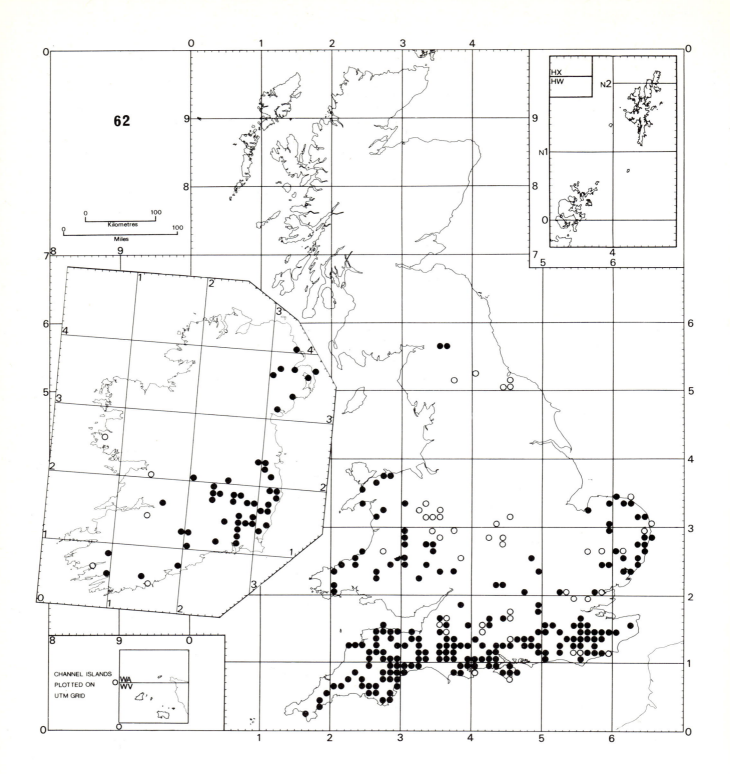

62 *Lecanactis premnea* (Ach.) Arnold

This is the dominant species of the *Lecanactidetum premneae,* and of the post-climax community of very old *Quercus* in ancient lowland parkland and woodlands of long continuity. It often covers large areas of rough bark usually in dry sheltered aspects low down on the bole. It is still widespread, though very local, in suitable habitats in southern England and Wales, but has largely disappeared from the Midlands and north Yorkshire mainly as a result of tree-felling and pollution by inorganic fertilizers and sulphur dioxide. Widespread in similar situations in temperate Europe where it is now extremely rare; distribution elsewhere uncertain.

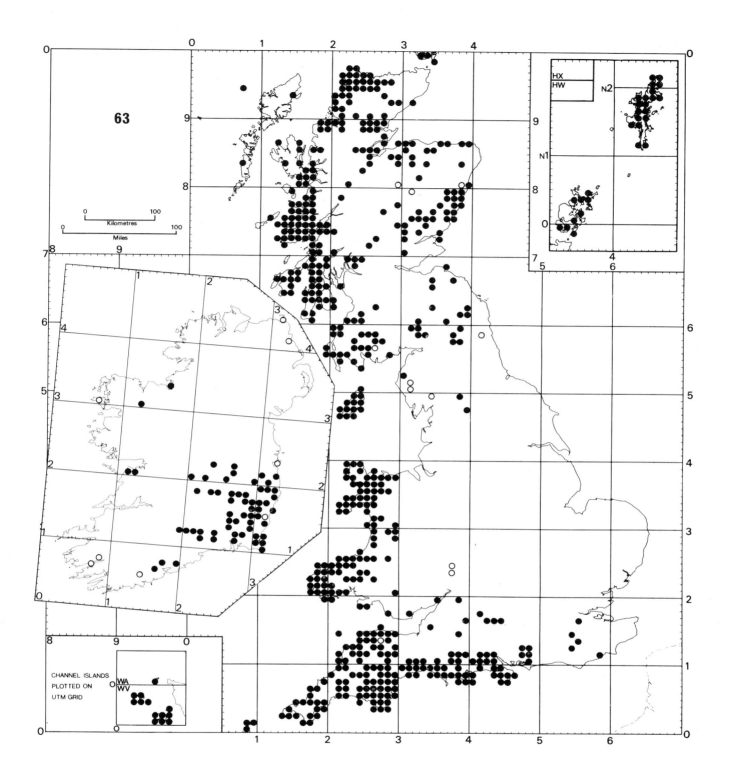

63 *Lecanora confusa* Almb.

A widespread species occurring on the smooth or sometimes rough bark of a wide range of young or old phorophytes including conifers and stems of *Calluna* and occasionally on old fence rails. Especially abundant in coastal areas, and a member of the *Lecanorion subfuscae,* a pioneer community of well-lit twigs and branches of hedgerow shrubs and trees. It is also moderately shade-tolerant, entering a facies of the *Graphidion scriptae* in young or medium-aged deciduous plantations. Probably scattered along the western seaboard of Europe as far as southern Sweden, but total distribution uncertain due to confusion with other species.

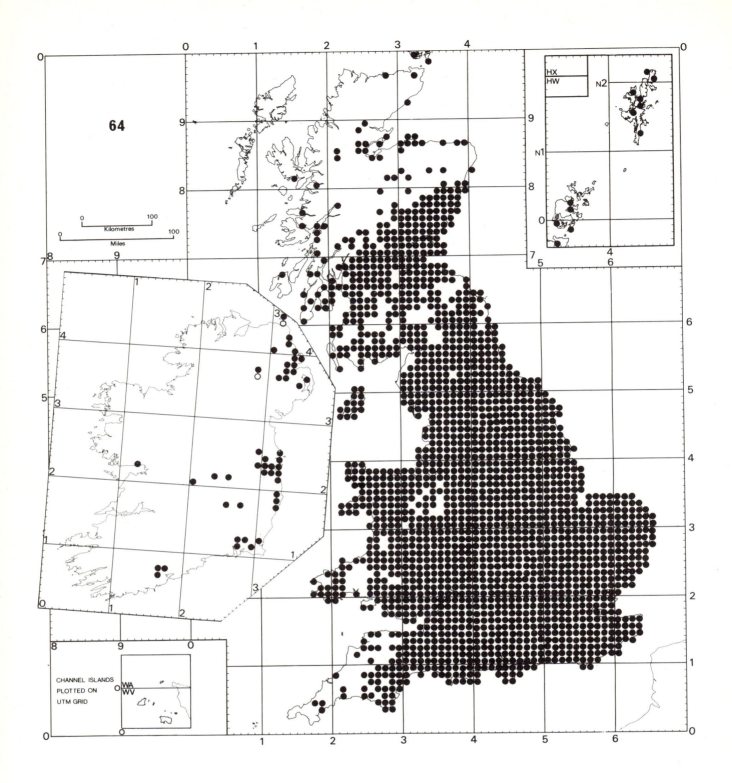

64   *Lecanora conizaeoides* Nyl. ex Crombie

A now ubiquitous species which, since 1860, has spread rapidly over large areas of Britain with a winter sulphur dioxide level of 50-150 μg m^{-3}. In urban areas it covers the boles, branches and twigs, and sometimes leaves, of most phorophytes where there is sufficient illumination and moisture. It also occurs on rocks, commonly sandstone and millstone grit, and brick as well as earth banks, especially in heathland areas. *L. conizaeoides* is often the exclusive member of a notably species-poor association, the *Lecanoretrum pityreae*. The species becomes rare north of the central lowlands of Scotland and in some western areas of England and Wales; in such areas it is confined to worked palings or twigs. Throughout polluted areas of lowland Europe, otherwise known only from Iceland and locally in North America (where it appears to be introduced).

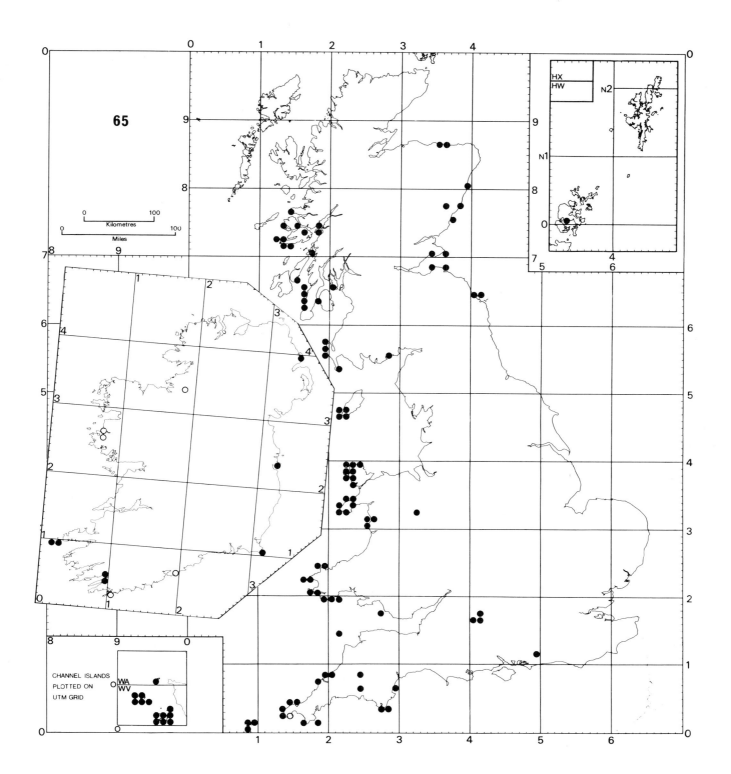

65 *Lecanora fugiens* Nyl.

A frequent and widely distributed coastal species, occurring in shallow fissures in hard rocks on exposed to moderately sheltered shores. Although chiefly confined to the xeric-supralittoral zones, the species is known from a few inland stations, where it grows on substrates such as standing stones (Stonehenge and Avebury, Wiltshire) and church walls. In the west it is widespread on all rocky coasts as far north as the Ardnamurchan Peninsula and at a few scattered stations in eastern Scotland and northern England; it is probably frequent on all suitable shores in Ireland. *L. fugiens* is easily distinguished from *L. dispersa* by the C+ orange, PD+ red reaction of the thallus and apothecia in the former. It is also known from western Norway and France.

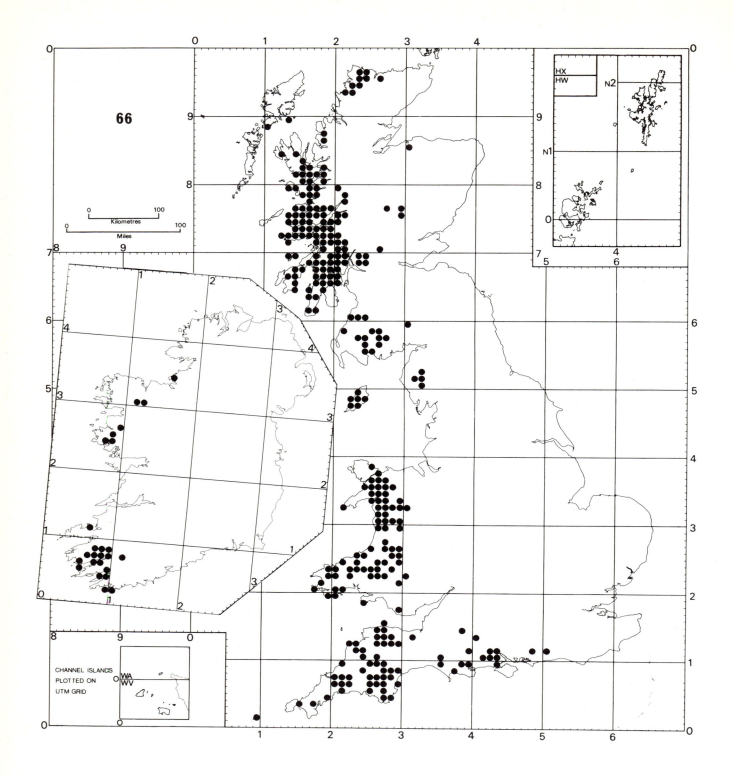

66 *Lecanora jamesii* Laundon

A local species of smooth boles and branches of a wide variety of phorophytes, especially *Salix,* but also *Fraxinus, Corylus, Fagus, Quercus, Ilex, Sorbus aucuparia,* etc. An important member of a sheltered, moisture-loving facies of the *Lecanoretum subfuscae,* forming scattered to confluent colonies of neat thalli with conspicuous rounded yellow-green soralia. In southern England it is rare and confined to damp, poorly-drained areas within old woodlands; in western areas it becomes more widespread and is frequent not only in moist woodlands, but also on *Salix* in carrs and by streamsides, where it often grows with *Lecidea carrollii.* Its absence from central and eastern England is due to drainage of suitable sites as well as air pollution. It is also known from western Norway, France (Brittany) and two stations in sheltered valleys in Austria.

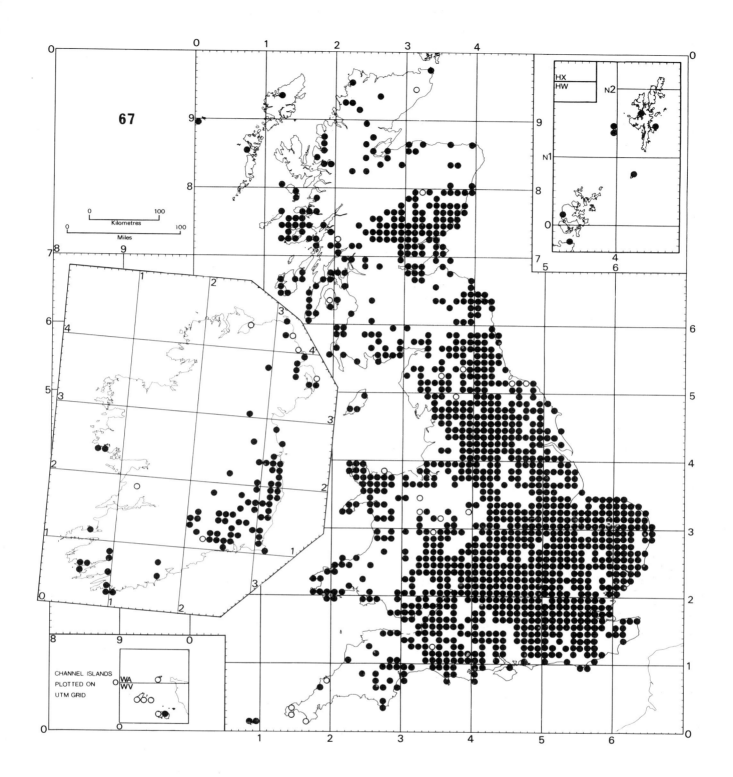

67   *Lecanora muralis* (Schreber) Rabenh.

Widespread on rocks and building materials, but occasionally corticolous, lignicolous or muscicolous; found in sunny and exposed situations, rare on steep slopes. Nitrophilous, preferring stonework frequented by birds; widespread in natural situations but limited to calcareous substrates in urban centres. Entering many saxicolous assemblages; synanthropic, on the increase throughout industrial, urban and lowland areas of the British Isles; usually absent where mean sulphur dioxide levels exceed 240 μg m^{-3}. Worldwide distribution, mainly in temperate and northern latitudes.

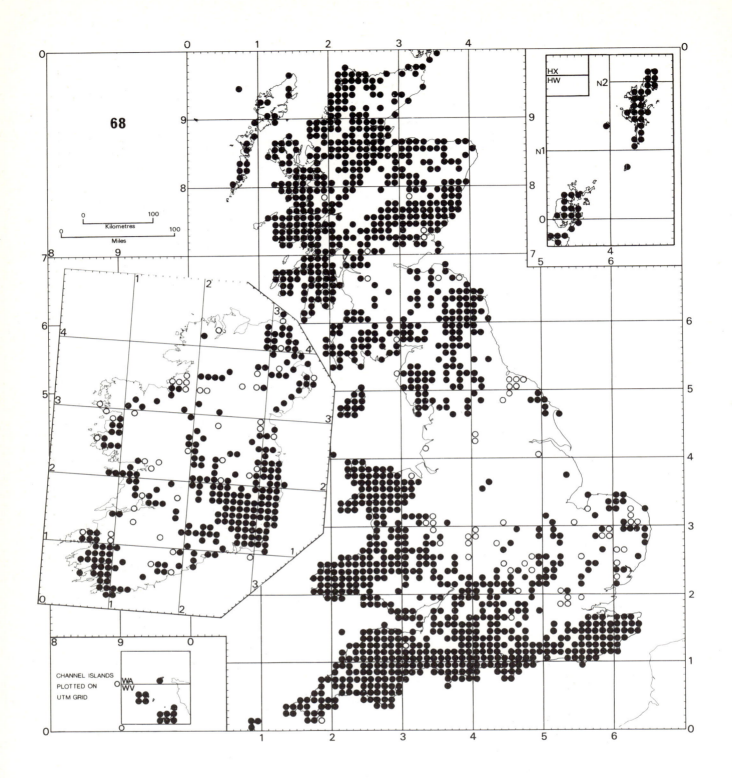

68 *Lecidella elaeochroma* (Ach.) M. Choisy

A very widespread, often abundant, mosaic-forming species of smooth, more or less nutrient-rich bark of a wide variety of young or old deciduous trees. It has a wide ecological amplitude and is a characteristic species of all facies of the *Lecanoretum subfuscae,* also entering adjacent facies of the *Graphidion scriptae* and *Xanthorion parietinae.* In non-polluted areas it is not only common on wayside trees and hedges, especially near the coast, but also on branches and boles of young or medium aged trees in plantations. Tolerant of concentrations of up to 100 μg m⁻³ of sulphur dioxide; at higher levels it survives low down on the boles of basic-barked trees, such as *Ulmus, Fraxinus* and *Acer pseudoplatanus.* A sorediate form (f. *soralifera* (Ericksen) D. Hawksw.) occurs mainly in the south and west. Widespread in Europe, distribution elsewhere uncertain.

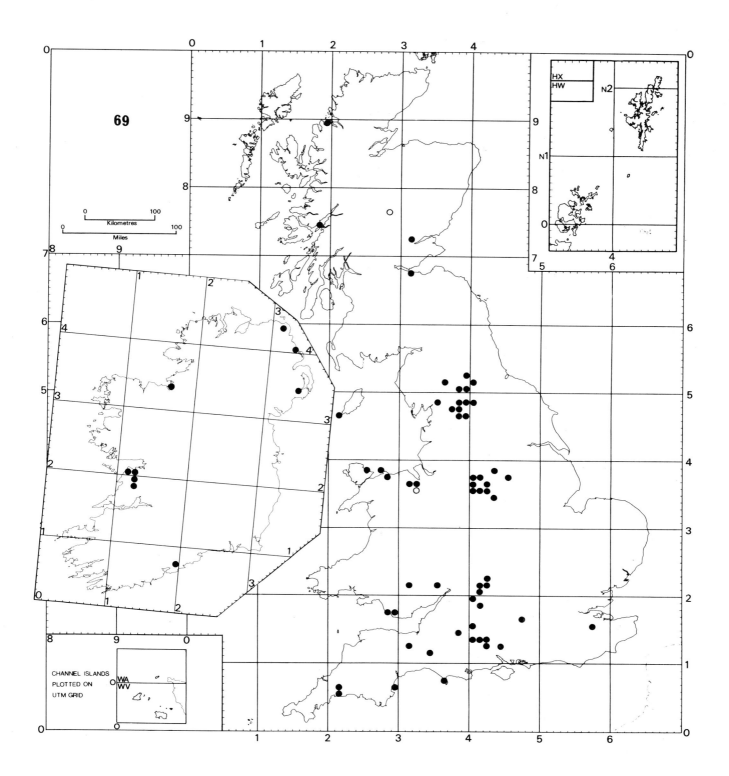

69 *Leproplaca xantholyta* (Nyl.) Hue

A characteristic species of pure limestones, belonging to the *Leproplacetum chrysodetae*. The species is restricted to moderately shaded, dry underhangs, recesses, cave entrances and, rarely, the sheltered sides of mortar-stone walls and limestone gravestones. The substrate is never directly wetted by rain or run-off. In England *L. xantholyta* is present, often sparsely, in most major exposures of pure limestone, especially Ingleborough (Yorkshire), Derbyshire and the Cotswolds. It is rare in Scotland and probably widespread in Ireland. A southern European species extending eastwards to Israel; distribution elsewhere uncertain.

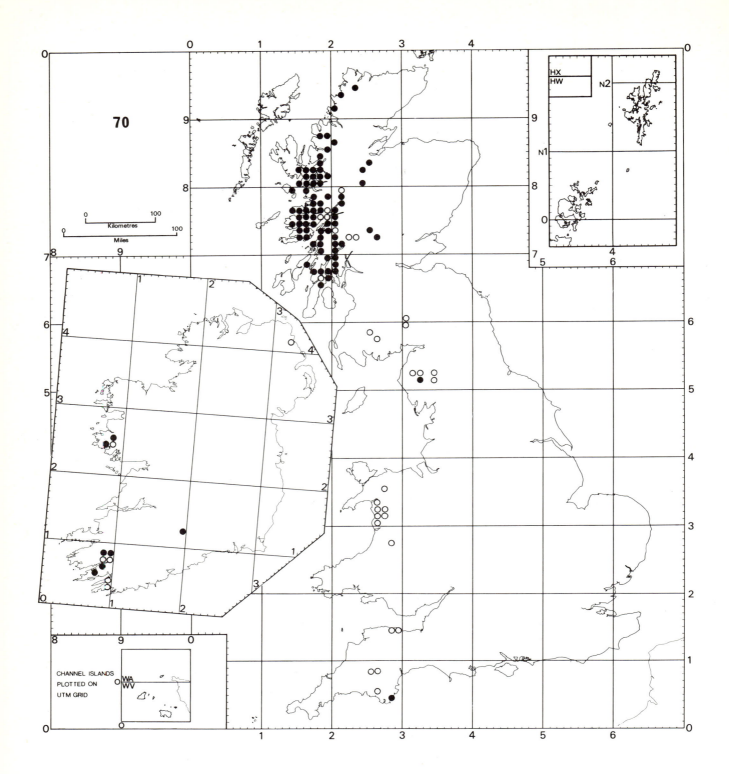

70   *Leptogium burgessii* (L.) Mont.

A species of very sheltered, humid valleys and streamsides with a high annual rainfall (exceeding 200 cm). Frequently associated with trees, such as *Corylus, Fraxinus, Salix* and *Ulmus,* or more rarely rocks, associated with limestone, basic volcanic or rich alluvial soils. *L. burgessii* is a member of a distinct facies of the *Lobarion pulmonariae* on substrates of pH 5.2-6.2 dominated by species with blue-green phycobionts. The species is now extinct in many of its old habitats in England and Wales where drainage and land reclamation, farming techniques and forest felling have taken a major toll; it is still abundant in suitable sites on the mainland of Scotland and the Inner Hebrides and in south and west Ireland. An extreme Atlantic species in Europe, extending from Portugal to south-west Norway.

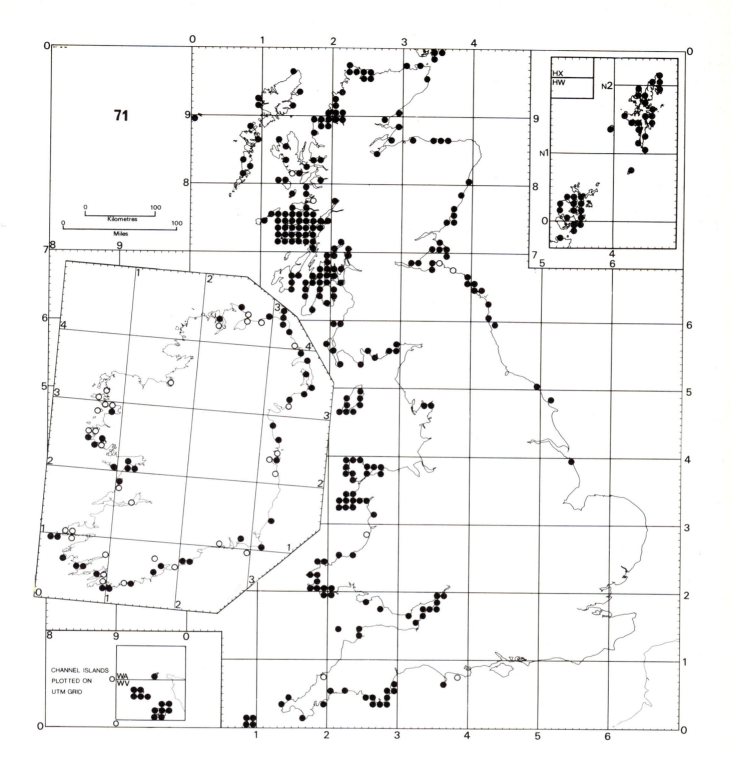

71   *Lichina confinis* (Müll.) Agardh

Found on rocks of all types, but avoiding excessively hard or smooth siliceous varieties. A very common lichen of seashores, straddling the interface between the littoral and mesic supralittoral zones. It is commonest on sheltered shores and disappears with increased wind and wave action. It is also rare on shaded or north-facing shores and is best developed in full sun. The plant often forms extensive swards and is normally fertile. This species may possibly be disappearing from shores in north-east England due to deposition of marine or estuarine silt. It has been fairly well recorded, being difficult to confuse with any other lichen except *L. pygmaea* (map 72), but is often mistaken for a small seaweed. *L. confinis* is widely distributed throughout the cool temperate and polar regions where its ecology is similar to that in Britain.

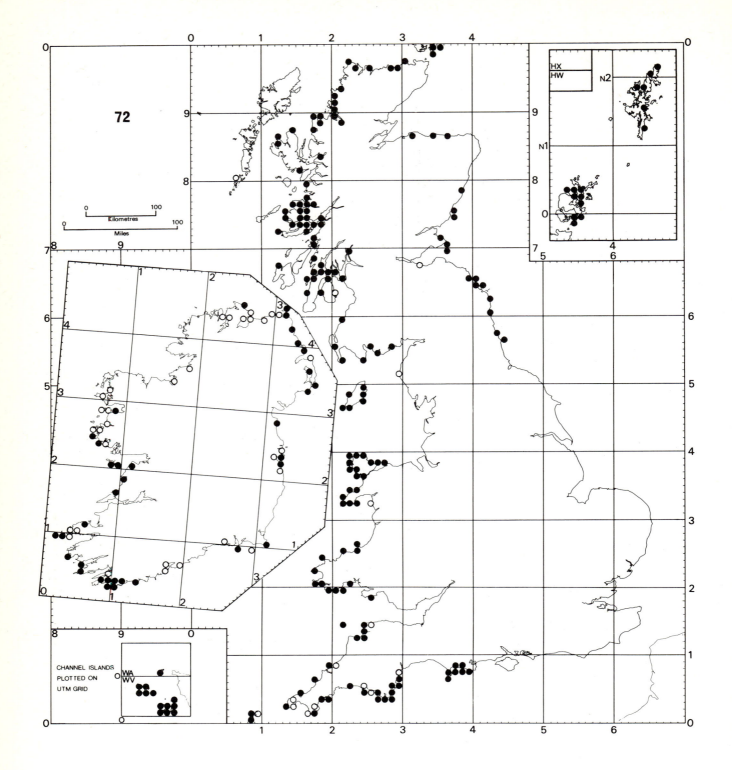

72 *Lichina pygmaea* (Lightf.) Agardh

Very common on rocks of all types provided they are neither excessively hard or smooth. Exclusively on seashores in the littoral zone, and especially common in western Britain; it descends further down the shore than any other British lichen and appears to need regular tidal inundation. Found only on moderately to fully exposed shores in full sun and absent from shade and shelter. Always fertile and often present as extensive swards amongst barnacles and fucoid algae; it provides food and shelter for many marine animals. The species may have disappeared from north-east England and estuaries due to silting. It has been fairly well recorded and is difficult to confuse with any other lichen, though easily mistaken for a seaweed such as *Catinella.* This lichen is found from the Azores to south-west Norway in similar habitats to those in Britain. The southern hemisphere var. *intermedia* is a member of the *L. confinis* group.

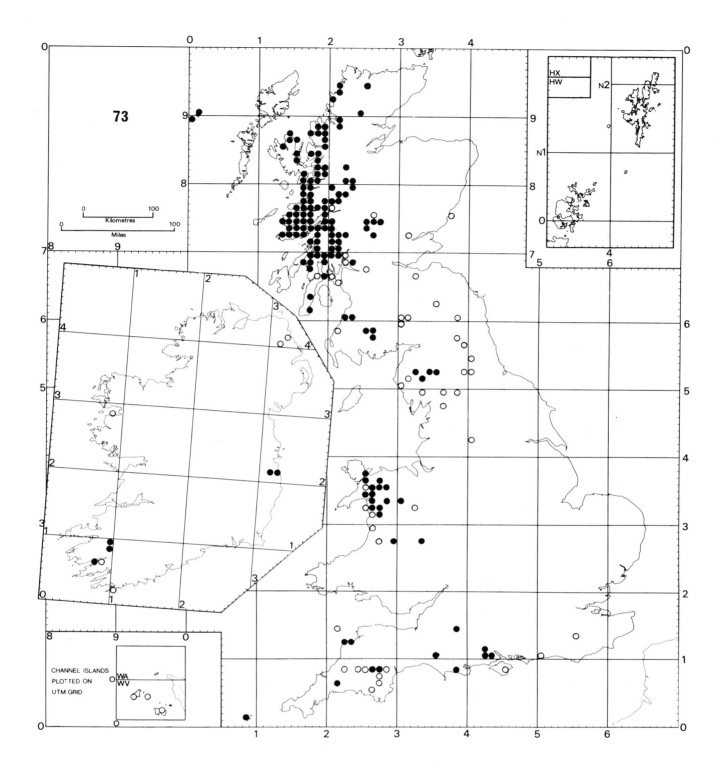

73 *Lobaria amplissima* (Scop.) Forss.

Locally frequent in the west Highlands, becoming rare and scattered in the central Highlands and south-west Scotland. Now very local in the English Lake District but still plentiful in a few sites in North Wales. Probably frequent in south-west Ireland, rare and widely scattered elsewhere. A typical member of the *Lobarion pulmonariae* but requiring more light than *L. laetevirens* (map 74) and *L. scrobiculata* (map 76) and hence more likely to occur, with *L. pulmonaria* (map 75), in tree crowns in high forest and on sunny boles at the margins of woods or isolated parkland trees. In Scotland it sometimes occurs on rather soft rocks in sheltered woodland sites near the sea. Chiefly associated with *Fraxinus, Ulmus, Acer pseudoplatanus* and *Quercus* with moderately basic barks (pH of over 5.0), it is an old woodland indicator species confined, except for western Scottish sites, to ancient woodlands and parks. Markedly pollution-sensitive and now absent from areas with over 35 μg m^{-3} SO$_2$. It has declined over most of its range, except in Scotland: this is due to pollution from SO$_2$ and farm sprays, felling and forest management of old woods and parks, and a lowering of the water-table. In Europe, it is an oceanic species common in relatively cool wet coastal sites, extending northwards to north Norway and southern Sweden; also reported from west France, Spain and Portugal.

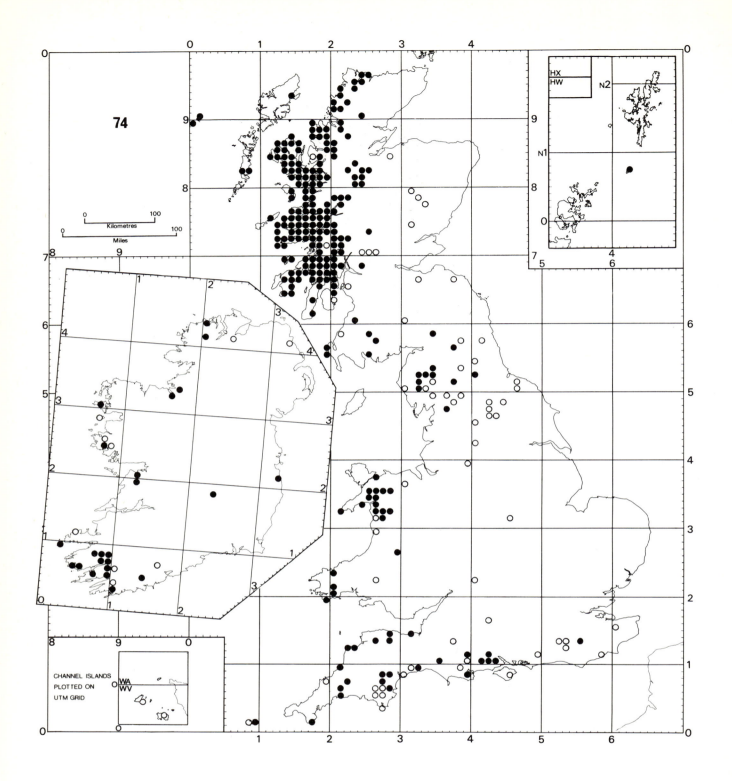

74 *Lobaria laetevirens* (Lightf.) Zahlbr.

A characteristic member of the *Lobarion* community, confined to ancient woodlands (except in a few western Scottish sites) and an old woodland indicator species. Mainly on *Quercus, Fraxinus, Ulmus* and *Fagus,* and other trees with moderately basic bark (pH over 5.0); pollution intolerant (not in areas with over 35 μg m⁻³ mean winter levels of SO₂); more shade-tolerant than other *Lobaria* spp. but requires more humid air. It is nearly always fertile. It has declined considerably, especially in the east where it has become extinct with destruction of old forest areas and lowered humidity, but much less than *L. scrobiculata* and *L. amplissima.* Also occasionally found on sheltered rocks in woods and on western sea-cliffs. It is still common in the west and central Highlands of Scotland, and occasional to locally frequent in western Britain and southwards to Cornwall, Dorset and the New Forest. In Ireland, except for the west where it is locally frequent, it is very rare. Widespread but very local in western Europe, from Norway to Spain but extinct in many lowland areas (e.g. Denmark, Germany); not extending very far east of the Pyrenees and central Europe now, and more oceanic than other *Lobaria* spp.

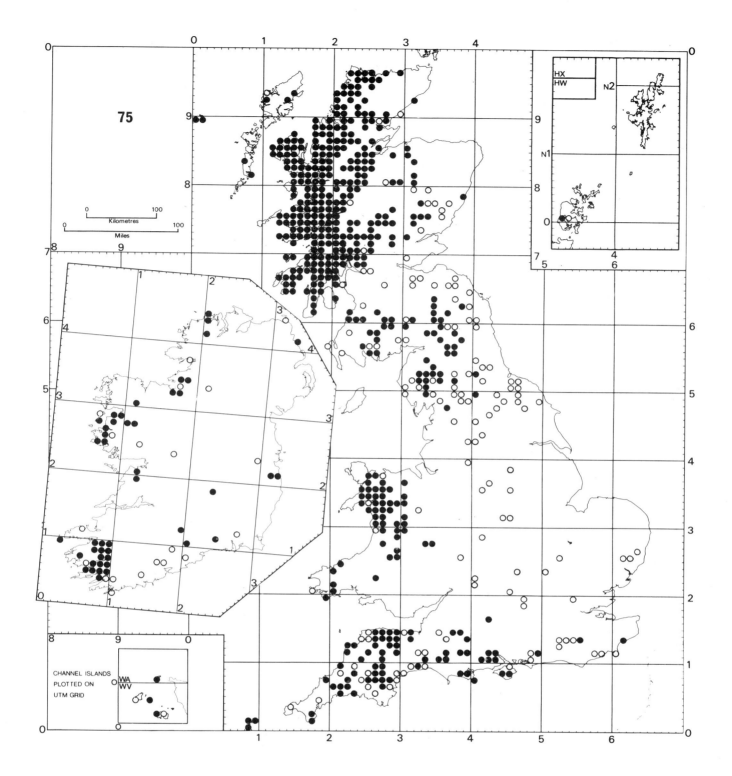

75 *Lobaria pulmonaria* (L.) Hoffm.

The most typical member of the *Lobarion* and (outside north-west Scotland, where it is still a colonising species) an excellent old woodland indicator. On many species of broad-leaved trees, but very rare on *Alnus*. Very pollution-sensitive (not in areas over 35 μg m^{-3} mean winter levels of SO_2) and, in areas of moderate pollution, more or less confined to more basic bark (pH 5.5) of old *Fraxinus Acer, Ulmus,* etc. Apparently not seen on conifers in Britain, but common on native *Abies* in old European montane forests. More tolerant of dry conditions and high light levels than other *Lobaria* spp., often on old trees in ancient open parkland like *L. amplissima* (map 73), but also occurs on mossy rocks and on sheltered coastal rocks especially in north-west Britain. Rarely fertile now, except in west Scotland and north Wales. It is still common to abundant in the west and central Scottish Highlands and occasional to frequent in south Scotland, north-west England, north-west Wales, south-west England and eastwards to the New Forest, but rare to very rare in south Wales and the south-east (Kent, Sussex). In Ireland it is abundant in the south-west, but scattered and rare elsewhere. It was formerly generally distributed, but is now extinct in central and eastern England; there are unlocalised 19th century records from Lincolnshire, the Isle of Man, etc. It has declined greatly both spatially and in abundance in England and is of relict nature in many sites at the eastern edges of its range. Often very small-lobed in drier areas, it assumes a sub-fruticose, robust form in moist regions. Widespread in Europe, but of sub-oceanic-montane pattern; extinct over most of the northern European plain south of Schleswig and east to west Poland due to pollution and changes in forest management; in the Mediterranean zone confined to old forests in the hills, but still in some dry but ancient relict *Quercus* forests near the coast. Also found in north Africa, Asia and N. America.

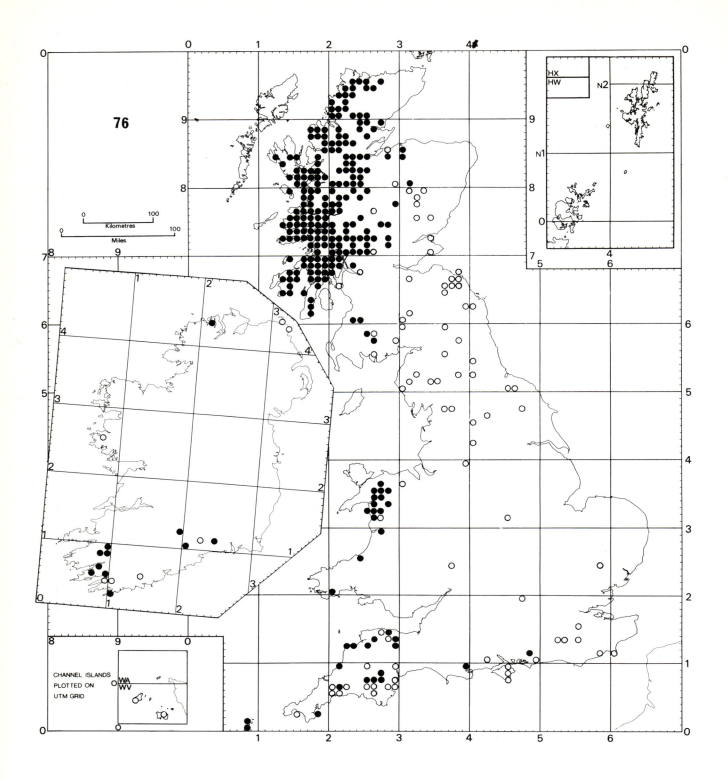

76 *Lobaria scrobiculata* (Scop.) DC.

A member of the *Lobarion* community, more tolerant of low humidity and high light intensity than *L. laetevirens* (map 74), extending further onto low pH substrates (down to pH 4.0) than other members of the genus, and penetrating more calcifuge associations, such as the *Parmelietum laevigatae*, in western Britain. An old woodland indicator species, found on various broad-leaved trees, but occasionally becoming saxicolous. One of the most sensitive of British lichens to air pollution (not now in areas with over *c.* 25 μg m^{-3} mean winter levels of SO$_2$), and hence has declined far more in England and lowland Scotland than other *Lobaria* spp. It is also much affected by farm pollutants. It is not now found fertile outside the western Highlands, but formerly it was fertile in Sussex (1805). It is common in the west, north and central Scottish Highlands, but in south-west Scotland, north Wales, south Wales, south-west England from Exeter westwards and southwards it is either rare or very local. Otherwise it is now extinct in England except for one Dorset site; extinct at the west Sussex site since *c.* 1975. In Ireland it is locally frequent in the south-west, but rare elsewhere (Donegal, Waterford). Widespread in Europe, but now extinct over most of the lowlands except in west France and in north Denmark (very rare); it is still frequent to locally abundant in montane forests (except where polluted) south to north Africa. The species is also known from Australia and New Zealand.

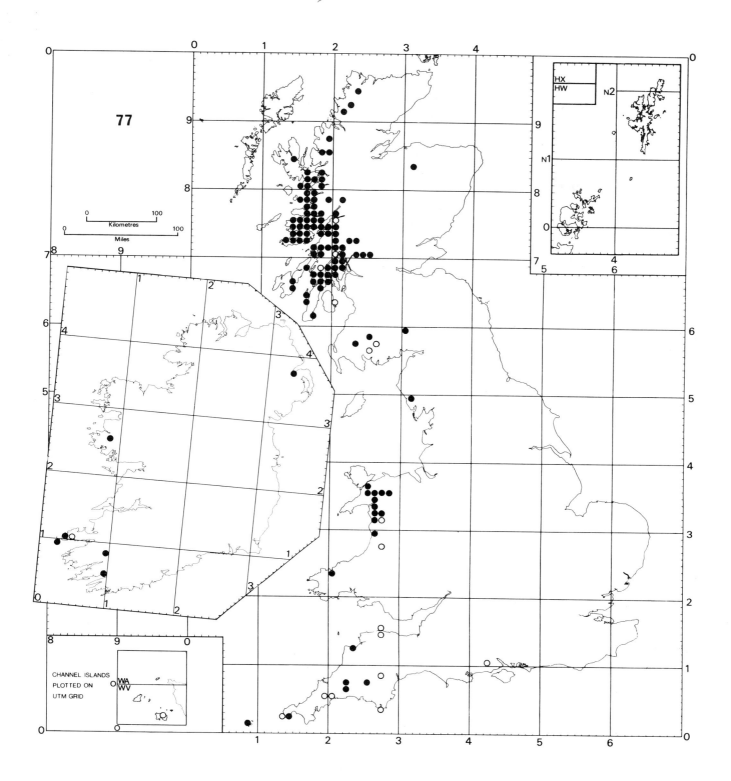

77 *Menegazzia terebrata* (Hoffm.) Massal.

This distinctive species mainly colonises the leached bark of *Betula, Alnus* and *Quercus,* mossy boulders in humid woods, often near streams, and north-east facing outcrops near or on the coast. Suitable habitats have a high annual rainfall of 120-230 cm and *c.* 180 wet days, supplemented in some stations (e.g. Isles of Scilly) by frequent on-shore sea mists. *M. terebrata* belongs to a lowland facies of the *Parmelion laevigatae* on substrates with a pH of 3.6-4.6. Locally frequent in some sheltered well wooded areas of the western mainland and inner islands of Scotland as well as north Wales and probably western Ireland. Elsewhere it is a very local species confined, often in small quantity, to a few undrained sites and rocky coastal areas. It is a rare hyperoceanic species in western Europe, occurring in northern Spain, Portugal, France (Brittany) and in mountain forests of Scandinavia; in central and eastern Europe it is a high montane *Fagus-Abies* species, only occurring in a continental association of the *Lobarion pulmonariae.* It also occurs in the Azores, north and central America (including Mexico), China and Japan; *M. terebrata* is unknown from temperate areas of the southern hemisphere, where most species of the genus occur.

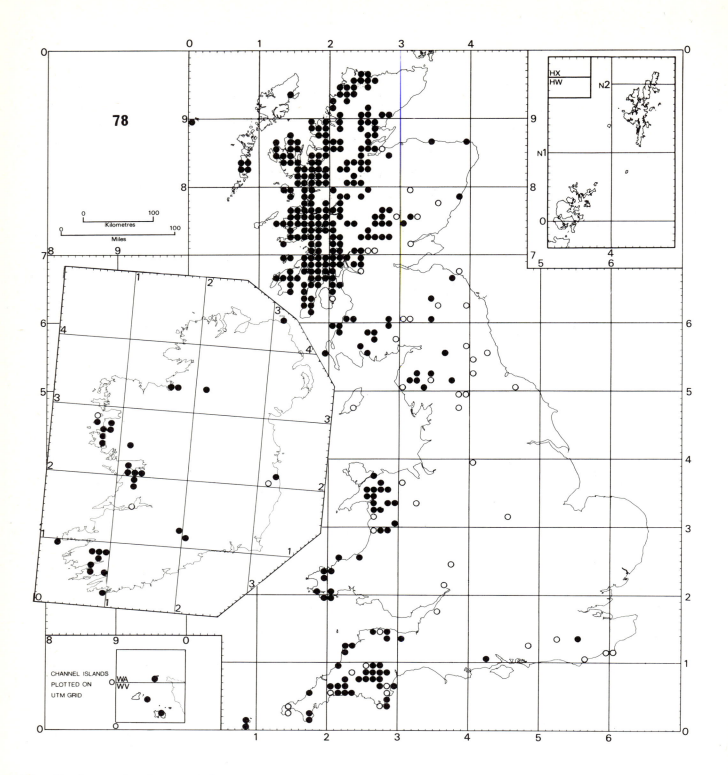

78 *Nephroma laevigatum* Ach.

An old woodland indicator species, mainly corticolous on *Quercus, Fraxinus, Corylus, Sorbus aucuparia* and *Ulmus glabra* in the British Isles, but also occurring on mossy rocks in open woodland and sometimes on sheltered coastal rocks. A characteristic member of the *Lobarion pulmonariae,* particularly of its more oceanic association, the *Nephrometum lusitanicae,* it occurs mainly in sheltered, but not very deeply shaded, lowland woodlands and in some ancient park-lands, either directly on bark or among bryophytes. Formerly probably more widespread in southern England, it is now confined to the western side of Britain, and is still very widespread and common in lowland valleys throughout much of Scotland. Still locally common in Ireland wherever relics of old woodland exist. It requires bark or rocks of moderately high base status (pH 5.0-6.5). It is normally fertile. Its decline in many areas is undoubtedly due to sulphur dioxide air pollution; it appears unable to tolerate levels above about 25 μg m^{-3}. Destruction or modification of ancient woodland (with removal of older trees, and lowered humidity due to opening out of woodland canopies and drainage) is an important factor in its decline generally. However, in more humid areas with very clean air (in west Ireland and west Scotland for example) it is still to be found colonising young trees, and even woodlands of relatively recent origin. Widespread in western Europe (from western Norway through Sweden and Denmark, west central and southern France, southern Germany to Portugal, Italy, Yugoslavia etc.), but now, through environmental change, largely extinct in the north-central European lowlands south of Jutland, and increasingly confined to the Atlantic coasts and montane regions in the interior. In southern and south-western Europe it is still common, but exclusively a species of montane forests, on both broad-leaved trees and on *Abies alba.* It is bicoastal in North America, and also known from north Africa and Macronesia.

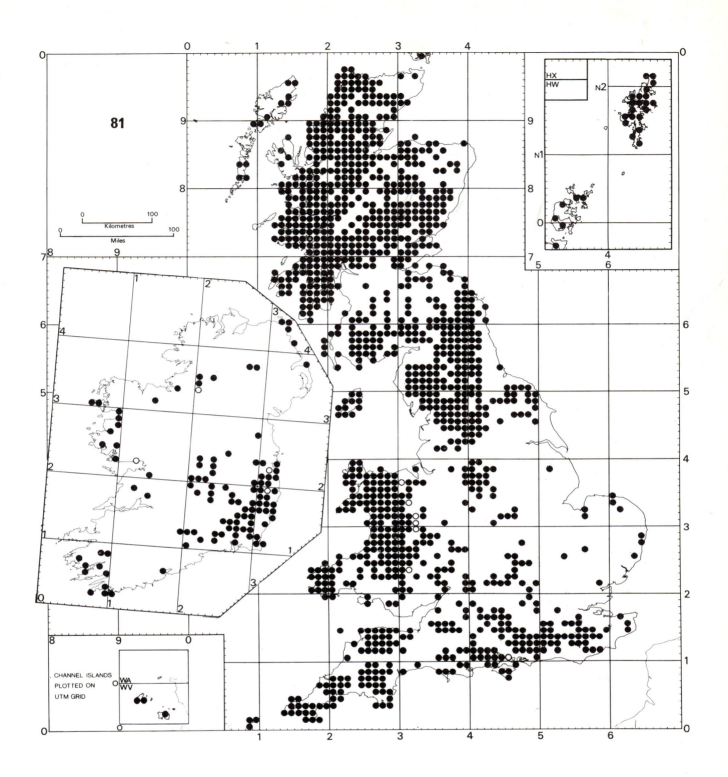

81 *Ochrolechia androgyna* (Hoffm.) Arnold

A species on smooth or rough, very acid bark of a wide variety of trees, *Calluna* stems, rock outcrops and boulders, at all altitudes from maritime sites (terrestrial halophilic) to 1340 m. Widespread in most relatively uncultivated moorland and mountain areas in western and northern Britain, where it occurs in the *Pseudevernion furfuraceae* and *Parmelion laevigatae* alliances and particularly in the *Arctoeto-Callunetum* of upland heaths. The species becomes local in eastern and central areas of Britain due to the effects of modern farming, tree-felling, reclamation of moorland and air pollution, but has probably increased in many areas with moderate pollution due to acidification of bark. A polymorphic species with creamy yellow soralia and containing gyrophoric acid. Abroad the species is circumboreal.

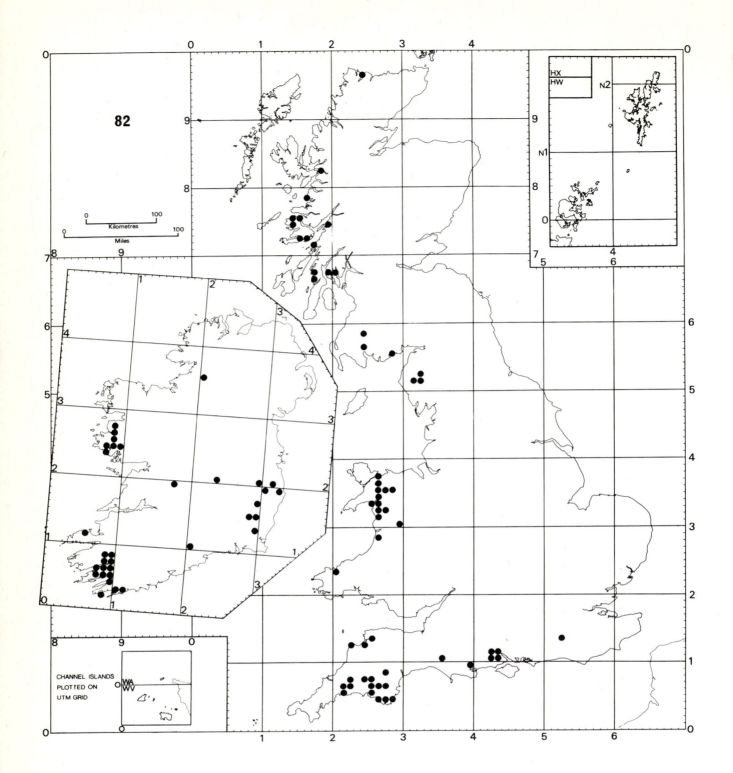

82 *Ochrolechia inversa* (Nyl.) Laundon

   A rare and local species of moist old woodlands, especially, but not exclusively, in sheltered, poorly-drained boggy situations, also sometimes on trees within wet carr. Mainly reported from old *Quercus, Fagus* and *Pinus;* very rarely saxicolous in west Scotland. Only locally frequent in a few southern and western districts of Britain; commoner in the old woodlands of the New Forest, and also in west, south-west and south-east Ireland. Sensitive to disturbance of old forest areas, and potentially also to air pollution, if this should spread to its strongholds. Occurs in an as yet undescribed association of the *Graphidion* in moist situations. Known outside the British Isles so far only from Rogaland, south-west Norway and the Azores.

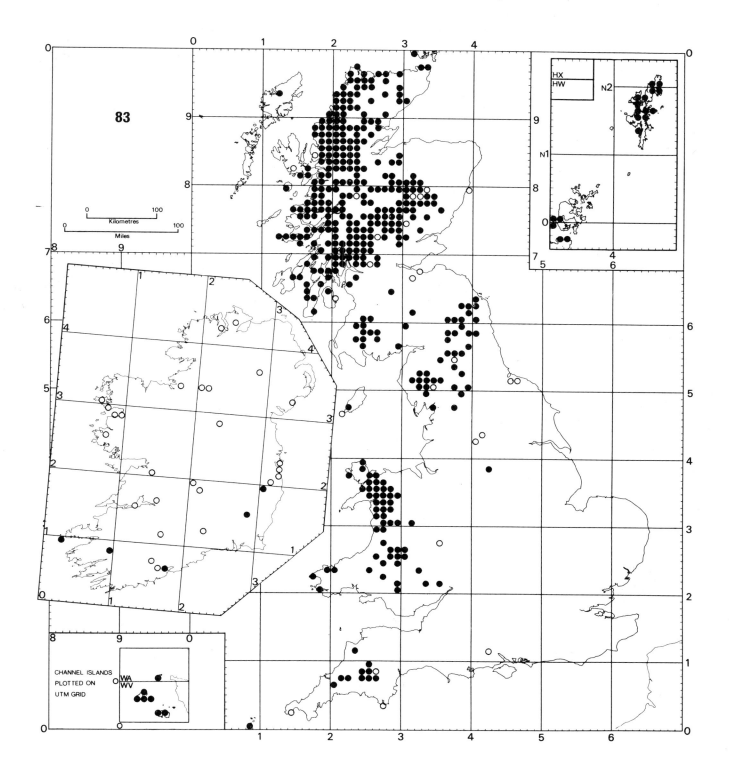

83 *Ochrolechia tartarea* (L.) Massal.

An upland species common in exposed wind-swept situations, in the Scottish Highlands in montane moss-lichen heaths, encrusting siliceous boulders where it may form extensive pure swards, and also on low pH (c. pH 3.5-4.5) bark in high rainfall areas, where it is an important constituent of the *Parmelietum laevigatae*. Locally abundant in northern England, Snowdonia and on high tors and in montane oakwoods on Dartmoor; also extending into the Peak District and probably Leicestershire. Formerly more frequent in the same areas where it was collected for dyeing wool. Circumboreal in the northern hemisphere, especially common in sub-boreal to arctic regions, but with oceanic tendencies; also in the mountains of central Europe and extending south to Corsica and Madeira.

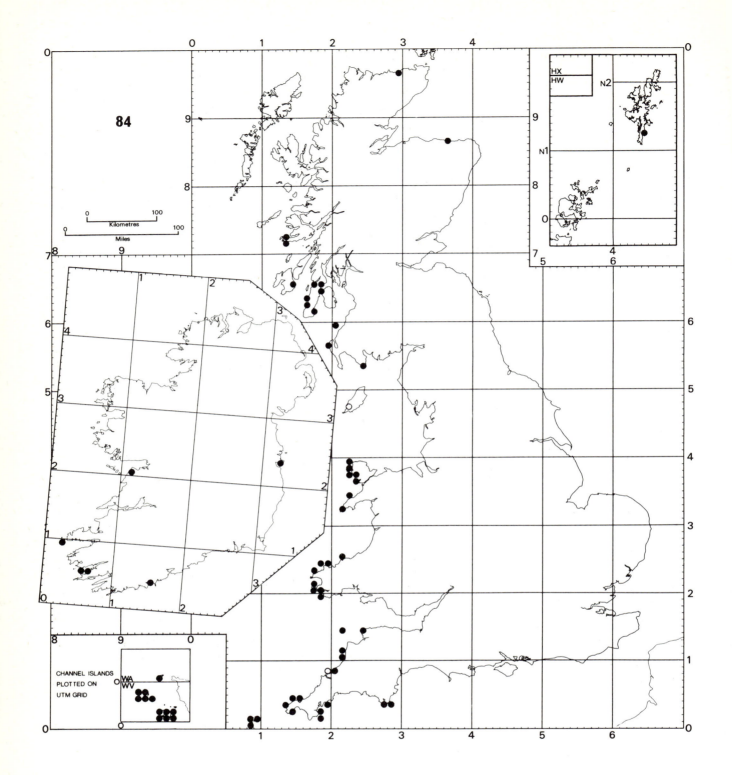

84 *Opegrapha cesareensis* Nyl.

A local species of dry, somewhat sheltered, recesses in siliceous maritime rocks not subject to direct rain. A member of the *Sclerophytetum circumscriptae,* characteristic of such situations; it is most frequent in the Isles of Scilly, Channel Islands and western Cornwall, but extends north to Caithness in particularly sheltered sites. Distribution elsewhere uncertain, but only otherwise reliably reported from western Germany; to be expected also in the Mediterranean.

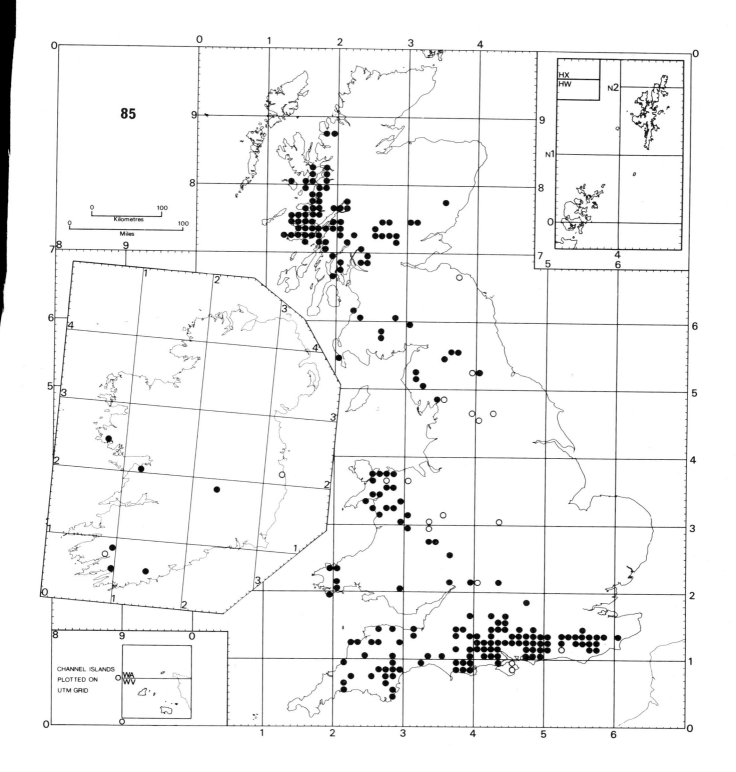

85 *Pachyphiale cornea* (With.) Poetsch

An old woodland indicator species, primarily of ancient *Quercus* trunks, but also on *Fagus*, *Acer campestre* and *Corylus* where it occurs intermixed with other species of the *Lobarion pulmonariae,* in sheltered, somewhat shaded, and generally humid situations usually directly on bark, though often among mosses. Formerly widespread throughout lowland Britain, but then probably often overlooked due to its small size. Now absent where mean sulphur dioxide levels exceed about 30 μg m^{-3}; this factor and the clearance of ancient woodland are responsible for its decline. Still frequent in ancient broad-leaved forests in parts of southern and north-west England, Wales, and western and central Scotland. Widely distributed in old woodlands in Europe, but becoming more or less confined to montane forests (600 m and above) southwards, and also known at least from North America and south Africa.

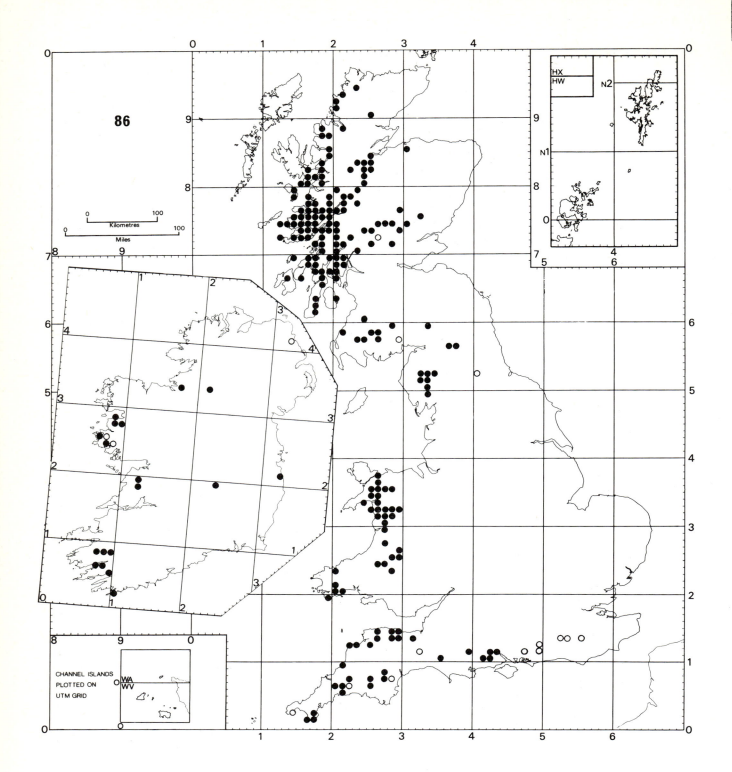

86 *Pannaria conoplea* (Ach.) Bory

A southern and western species occurring on mossy bark of broad-leaved trees in moderately shaded and continuously humid, sheltered woodland situations (exceptionally on rocks). A member of the *Lobarion pulmonariae,* it is often associated with other members of the *Pannariaceae* and its occurrence is adversely affected by both air pollution and woodland management involving an opening-up of the canopy. Locally abundant in north-western Scotland and western Ireland. Widespread in montane areas of Europe, but in the lowlands only near the Atlantic coast, from south Norway and western France to Portugal, and inland at Fontainebleau. Extinct in Denmark. Probably circumboreal, in cool moist temperate areas of the northern hemisphere, bicoastal in North America; also in montane situations in the tropics.

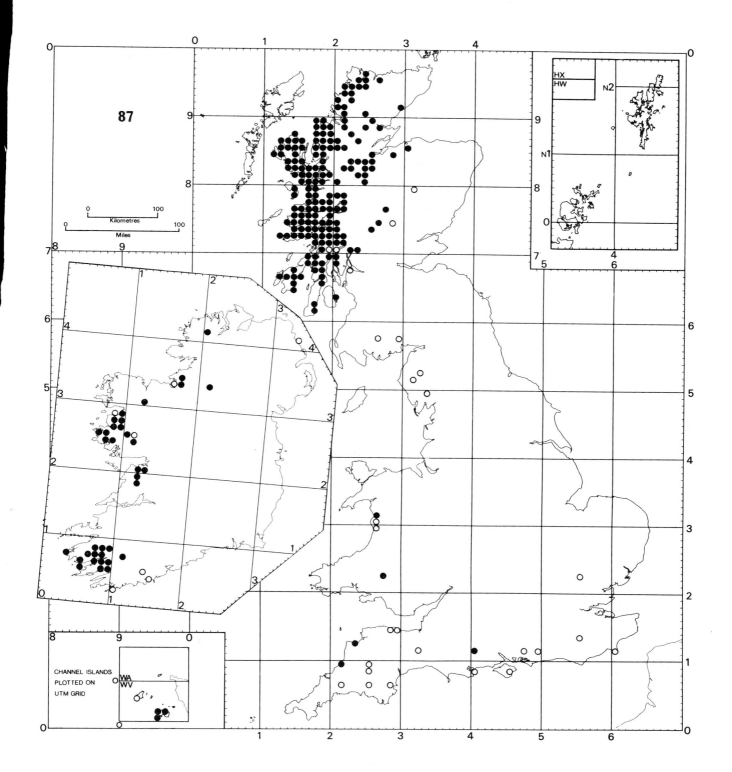

87 *Pannaria rubiginosa* (Ach.) Bory

An extreme oceanic species in Britain, widespread only in north-west Scotland and western Ireland where it occurs on mossy broad-leaved trees (exceptionally rocks) in mild and continuously humid, sheltered woods. It normally occurs in moderately shaded woodland interiors, more rarely in open sites in western Scotland, and, in common with other members of the *Lobarion pulmonariae,* has been lost from many English sites due both to the disturbance of woods with a long history of ecological continuity and to its extreme sensitivity to sulphur dioxide pollution. A widespread pantropical to temperate species occurring sporadically through Europe in suitable situations, in both highly oceanic lowland areas and humid montane regions in the south. Known from all continents except Australia and most of Oceania.

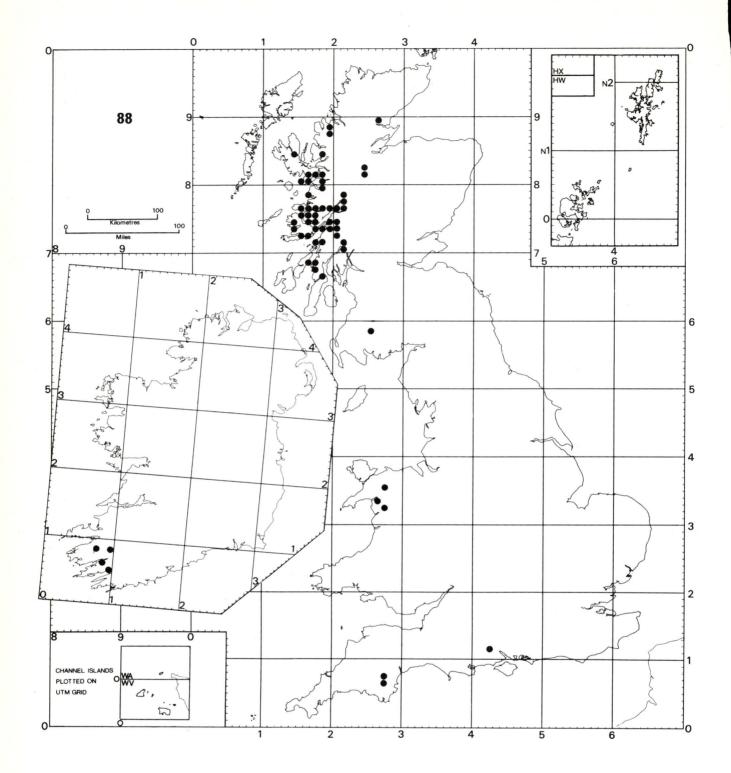

88 *Pannaria sampaiana* Tavares

A local southern and western species of continuously humid and warm woodlands where it occurs mainly on smooth bark of *Fagus sylvatica* and *Fraxinus excelsior,* and also over mosses (exceptionally on mossy rocks in Scotland and Scandinavia). A member of the *Lobarion pulmonariae,* it is often associated with other *Pannaria* species and like them is likely to be endangered by any woodland management leading to an opening-up of the canopy. Largely confined to extreme oceanic areas of Europe, in western Norway and from Brittany south to Portugal and Spain, and east to Italy and Yugoslavia. In south Europe it is especially associated with *Castanea* forests, in the mountains from 800 to 1200 m. Otherwise known only from Tunisia (a single collection).

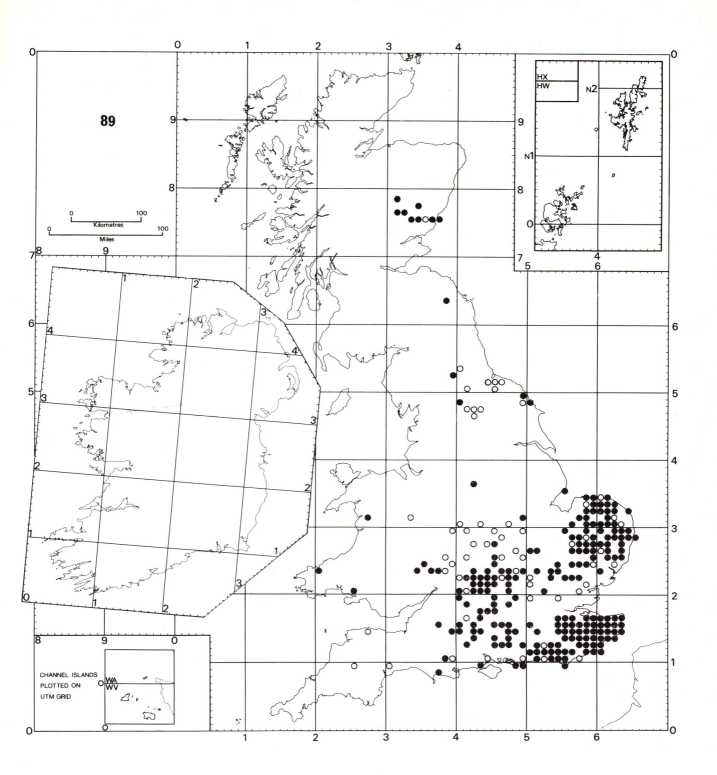

89 *Parmelia acetabulum* (Necker) Duby

A distinctly eastern species in the British Isles extending into eastern Scotland, but now much reduced in the north-central and north-eastern counties of England where mean sulphur dioxide levels exceed about 60 μg m⁻³ ; its western limit coincides well with the 1000 mm rainfall isohyet. It occurs on nutrient-rich bark, especially in field margins, parklands and along dusty roadsides, always in sunny situations and almost exclusively on trunks; an important component of the eastern facies of the *Physcietum ascendentis* in Britain. A southern and eastern species in Europe, extending northwards into central Sweden and southwards into North Africa. In southern Europe it has a wide altitudinal range as a forest species, occurring even on *Abies* at quite high altitudes. Perhaps endemic to Europe.

90 *Parmelia arnoldii* Du Rietz

A rare, strongly oceanic species occurring in moderately well-lit, mild and constantly humid broad-leaved woodlands, especially in valley bottoms, with a long history of ecological continuity. Usually amongst mosses on horizontal boughs of *Quercus* or *Salix* in very moist situations and associated with members of the euoceanic facies of the *Lobarion pulmonariae*. Its status is likely to be threatened by both increasing air pollution and disturbance of ancient woodland, the latter resulting in its loss from one of its two Devon sites in the last few years. Perhaps more frequent in western Ireland and Scotland than records imply, as it is easily passed over for the common *Parmelia perlata* (map 102). A warm-temperate species, occurring rarely in hyperoceanic areas of western Europe north to the extreme south of Norway, also in the northern Alps and Tyrol. Otherwise known only from Central, North and South America.

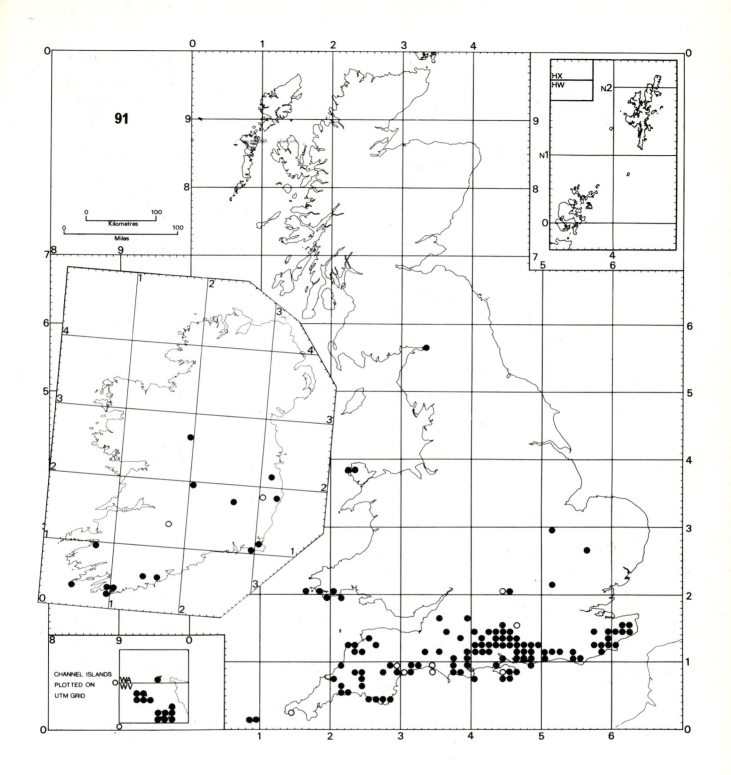

91   *Parmelia borreri* (Sm.)Turner

A species of well-lit, moderately nutrient-rich bark on the trunks, and particularly upper branches, of broad-leaved trees in sheltered sunny situations. Distinctly southern in the British Isles, but extending to the Solway Firth (the northernmost European site) where low rainfall and the level of sunshine permit. Especially characteristic of the *Parmelietum carporrhizantis* in South Devon, but a component of the *P. revolutae* over most of its range. Formerly much confused with *Parmelia subrudecta* (map 110) which is more rarely fertile, tan, not black, below and chemically distinct. Widespread in southern and western Europe, rarely in the Alps, absent from Scandinavia, common on the east and south African mountains, and also known in the U.S.A. (very rare).

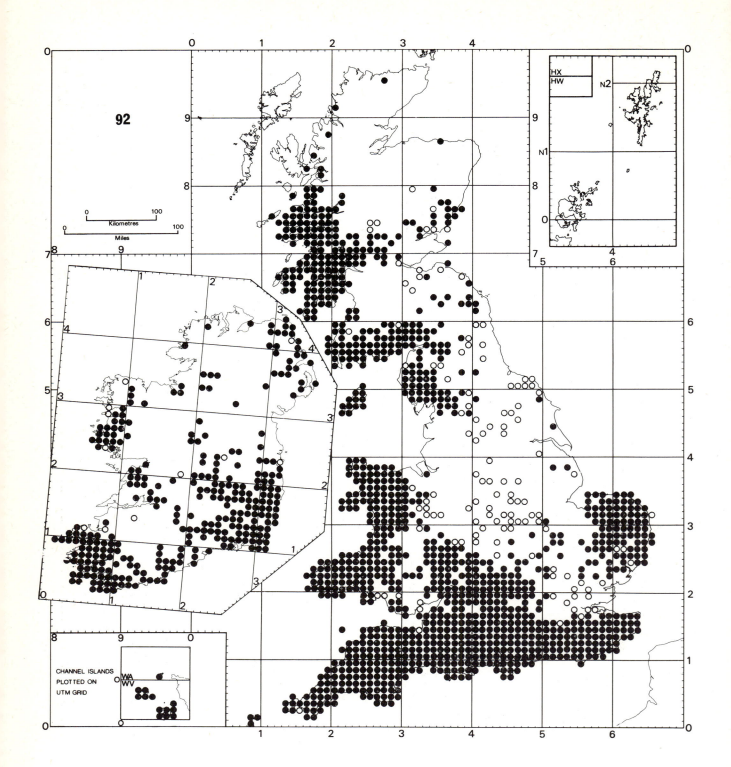

92 *Parmelia caperata* (L.) Ach.

Formerly common and widely distributed throughout lowland Britain, but now lost from large areas of central and north-eastern England where mean sulphur dioxide levels exceed about 55 μg m⁻³. One of the most important epiphytic lichens in unpolluted parts of lowland Britain, occurring on well-lit moderately acid (pH 5.0-5.5) broad-leaved trees as a major species of the *Parmelietum revolutae,* the predominant association on non-nutrient-enriched bark in the region. Occasionally also entering other communities, especially near the sea, and able to grow on coniferous trees and fences, siliceous rocks and walls, and in short coastal turf. Absent from large areas of central and northern Scotland and also parts of east-central Wales and north-east England, where the climate is either too low in sunshine or too cold in winter for it. A widely distributed, pantemperate species, present in all continents, but extending north only into south-west Norway (very rare). Much commoner in the southern half of Europe than the northern, and now much reduced across the northern European plain (from Belgium to Poland) by air pollution.

[Additional Scottish records added in proof: 18/14, 23, 24, 25, 26, 32, 33, 35, 41, 42, 43, 44, 45, 47, 51, 52, 53, 55, 56, 61, 71, 72].

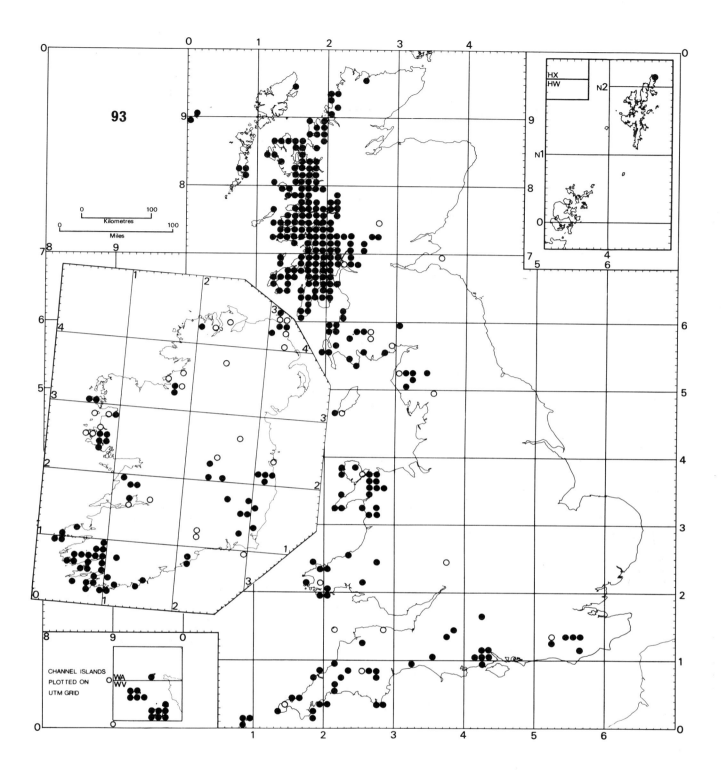

93 *Parmelia crinita* Ach.

A western and southern species sensitive to disturbance, and characteristic of sites with a long history of ecological continuity. It occurs mainly on the trunks and upper branches of broad-leaved trees in sheltered situations, and on exposed sunny coastal rocks, in south-west England and Wales. It occasionally enters into communities of the *Parmelietum laevigatae,* but is most characteristic of better-lit facies of the *Lobarion pulmonariae.* Probably formerly more widespread in lowland Britain, but adversely affected by air pollution. A pantemperate to tropical and sub-boreal species, occurring in Europe north to south-west Norway, also present in montane areas of southern, central and eastern Europe and in lowland forests on and near the Atlantic coast of western France; reported from all continents.

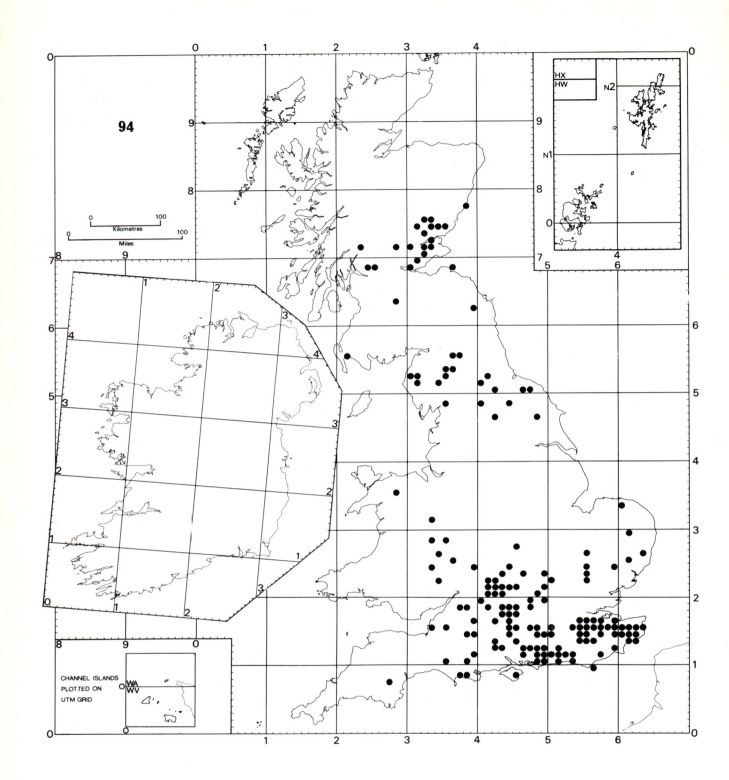

94 *Parmelia elegantula* (Zahlbr.) Szat.

A lowland eastern species in the British Isles, not collected here until 1965, and evidently extending its range. It occurs on moderately nutrient-rich bark of pH 3.8-4.6, and can withstand mean sulphur dioxide levels of up to about 80 μg m^{-3}; more rarely on tombstones and wall-tops. Often accompanied by *Parmelia laciniatula* (map 99), and with it forming a distinctive association, the *Parmelietum elegantulae*. Probably circumboreal in the northern hemisphere, from temperate to arctic zones. Characteristic of *Abies* forests in the high montane zones of central and southern Europe. Also present in north Africa, Pakistan and temperate South America.

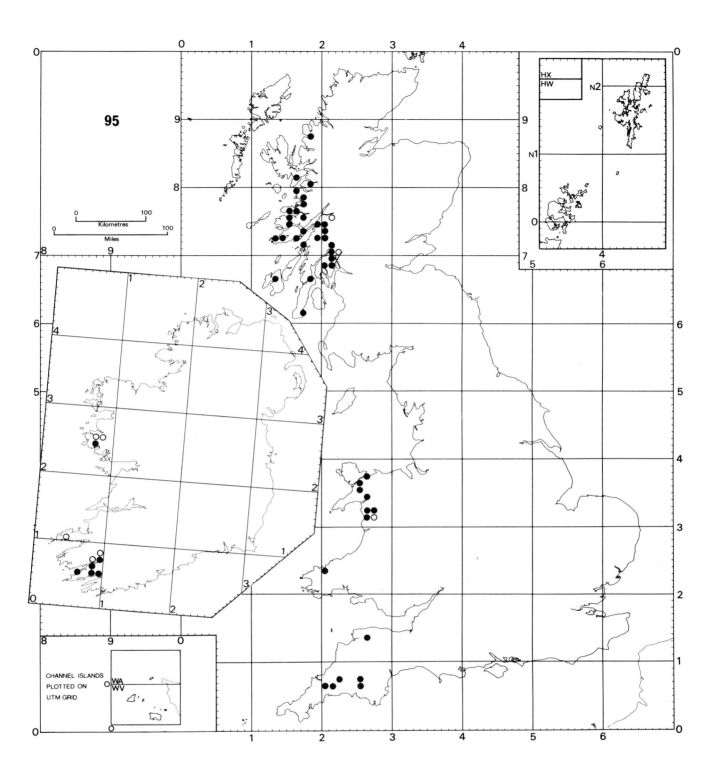

95 *Parmelia endochlora* Leighton

A markedly oceanic species in the British Isles, found in mild, constantly humid broad-leaved woodlands with over 180 wet days per annum, especially in sheltered valley carrs. It is a faithful, but very local, member of the *Parmelietum laevigatae,* but can sometimes be found in connection with the euoceanic facies of the *Lobarion pulmonariae.* It occurs on leached, moderately well-lit bark, often amongst mosses, and, rarely, in similar situations on mossy rocks (especially within woods). A strictly Atlantic species in Europe, extending northwards from Portugal to western France (with an isolated inland outlier at Fontainebleau) and the British Isles, which includes its most northern localities. Also known from humid forests in Central and South America, Madagascar, east and south Africa, the Azores, St. Helena and Hawaii.

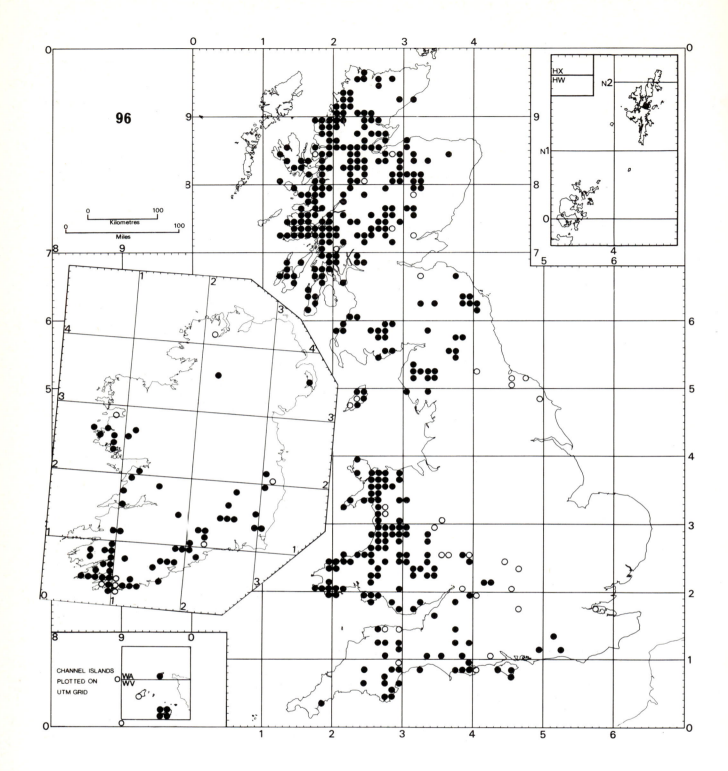

96 *Parmelia exasperata* de Not.

Probably formerly scattered throughout the British Isles, but now absent where mean sulphur dioxide levels exceed about 45 μg m⁻³. Almost entirely restricted to the *Lecanoretum subfuscae* pioneer community of moderately acidic twigs (sometimes also fence posts and exceptionally rocks) in well-lit, but somewhat sheltered sites. Typically occurring as isolated individuals, and so easily overlooked; commonly fertile. Widespread in unpolluted parts of Europe, and extending from northernmost Norway southwards through to north Africa and Macronesia; locally abundant in eastern North America, but otherwise only known from the Kazakh region of the U.S.S.R.

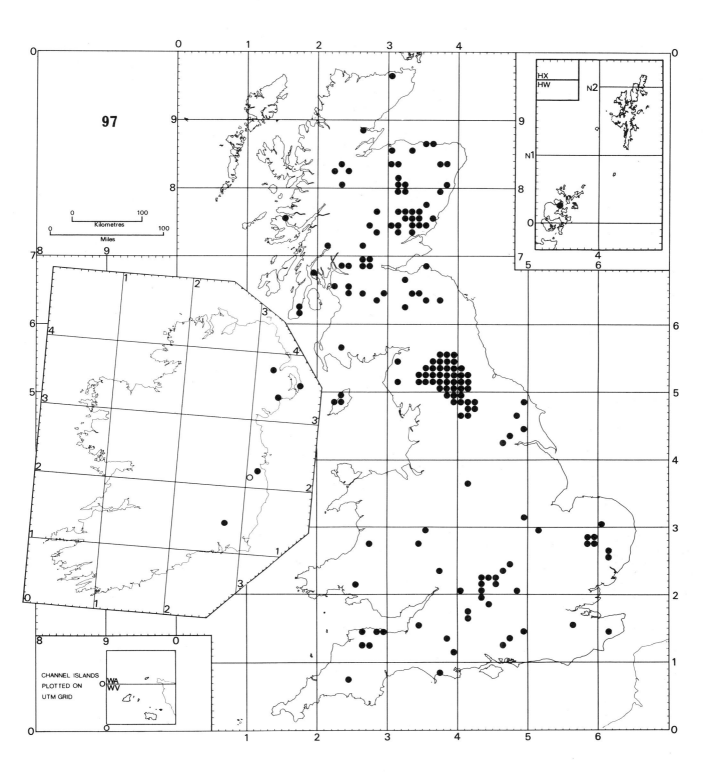

97 *Parmelia exasperatula* Nyl.

Now a widespread species in the British Isles, it was first collected here in 1888 and may be increasing; in Ireland, found so far only in the east. It is especially abundant in the northern Pennines, occurring on nutrient-rich bark and especially wall-tops below the canopies of roadside trees; also on birds' perches and, more rarely, tombstones. Usually present on nutrient-enriched, but acidic, substrates in an eastern facies of the *Physcietum ascendentis*. Moderately tolerant of sulphur dioxide air pollution (to *c.* 70 µg m⁻³) and able to exploit habitats from which other species have been eliminated. Probably circumboreal in the northern hemisphere, in Europe extending north from Portugal and the Pyrenees; widespread in temperate areas of North America, and also known from Pakistan and Siberia.

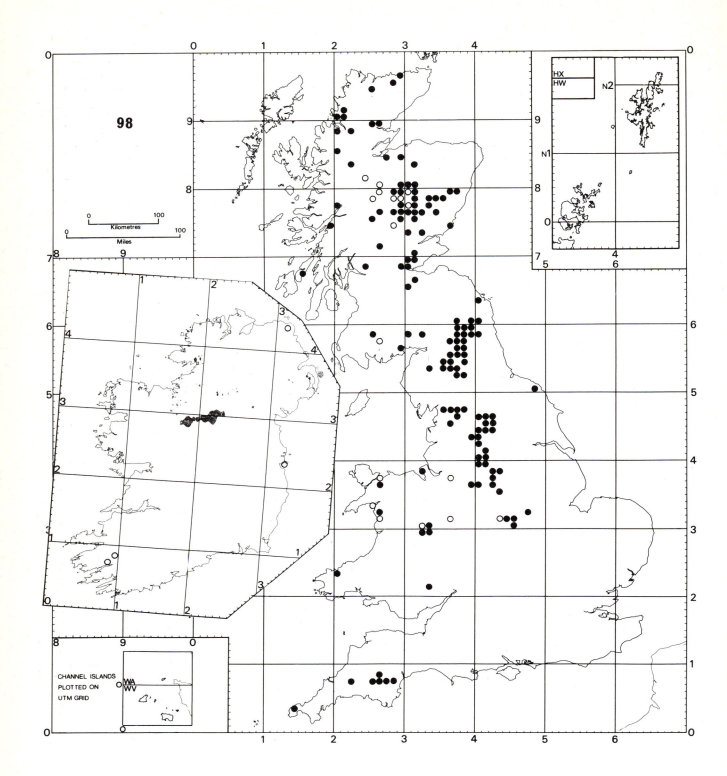

98 *Parmelia incurva* (Pers.) Fr.

A species of coarse-grained, well-lit siliceous rocks (exceptionally wood) in upland areas, typical of the arctic-alpine *Rhizocarpion alpicolae* in the Cairngorms, but also occurring occasionally in the *Umbilicarion cylindricae* in more lowland situations. Especially frequent in Charnwood Forest, the Peak District, and Pennine chain in the *Fuscideetum kochianae*. Apparently tolerant of mean SO_2 air pollution levels to *c.* 70 μg m^{-3}, and increasing in the above sites due to the elimination of competitors; also colonizing sandstone walls there. A boreal species apparently confined to Europe, extending south to the Alps (where it is very rare).

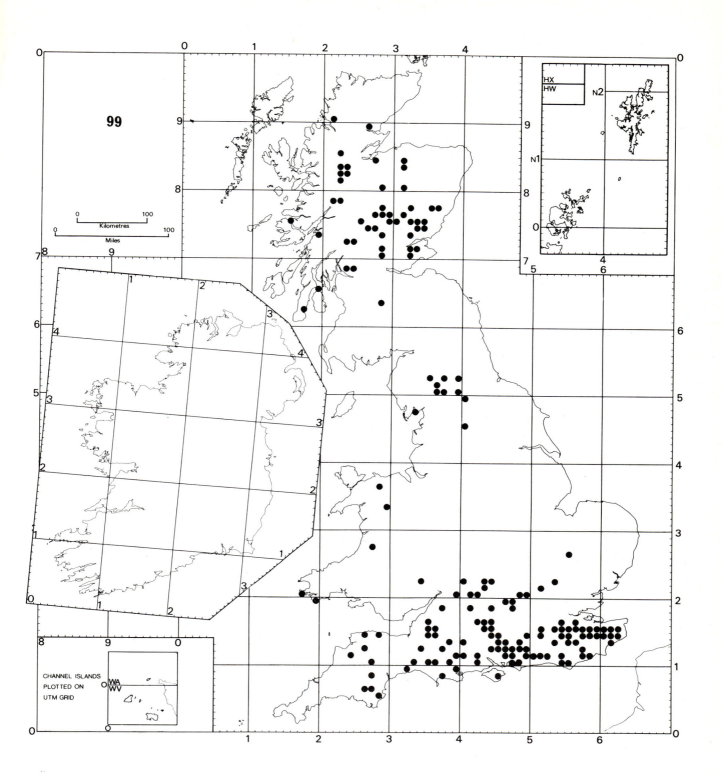

99 *Parmelia laciniatula* (Flagey ex H. Olivier) Zahlbr.

First collected in the British Isles in 1933, this species is commonest in south-east England, but has a scattered distribution as far north as Sutherland. It occurs mainly on moderately well-lit trunks and leafy twigs of a wide range of broad-leaved trees (with bark pH 3.6-4.8), but can occasionally be encountered on coniferous trees or siliceous rocks and walls. A moderate tolerance of sulphur dioxide air pollution (mean values to *c.* 65 μg m^{-3}) has enabled it to spread in areas where other species have been eliminated. Often growing with *Parmelia elegantula* (map 94) as part of the *Parmelietum elegantulae*. A European species, with somewhat continental tendencies, extending from Morocco into southern Scandinavia; in the south it is essentially a species of high montane *Abies* forests.

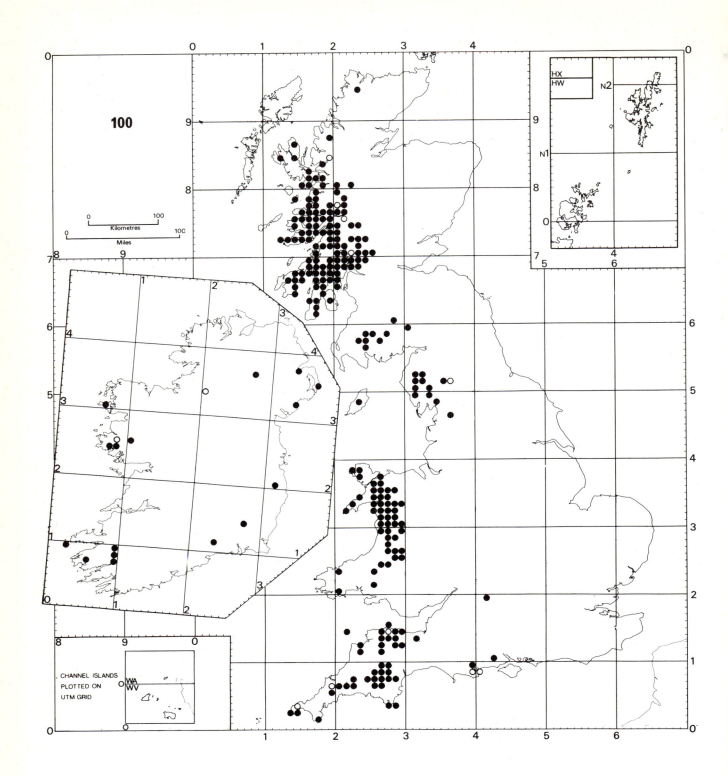

100 *Parmelia laevigata* (Sm.) Ach.

A hyperoceanic species of areas normally with an annual rainfall of at least 127 cm over 180 rain days, but also requiring cool summer temperatures. It occurs in well-lit oak or birch woodlands exposed to heavy rain which leaches the bark to pH 3.7-4.6. A characteristic and often dominant species in the *Parmelietum laevigatae,* it also occurs on mossy rocks in cool western valleys and coastal cliffs where the humidity levels are maintained (e.g. South Devon coast). Appears to be sensitive to sulphur dioxide, but apparently increasing locally in areas subjected to increasingly acid rainfall in western Scotland (Cowal). A strongly oceanic species in Europe, occurring only near the Atlantic seaboard from Normandy south-westwards, except for an isolated outlier at Fontainebleau in north-central France, and in southern Norway; present in humid situations in the northern Alps. Also known from the eastern U.S.A., Central and South America, Tasmania and the east African mountains.

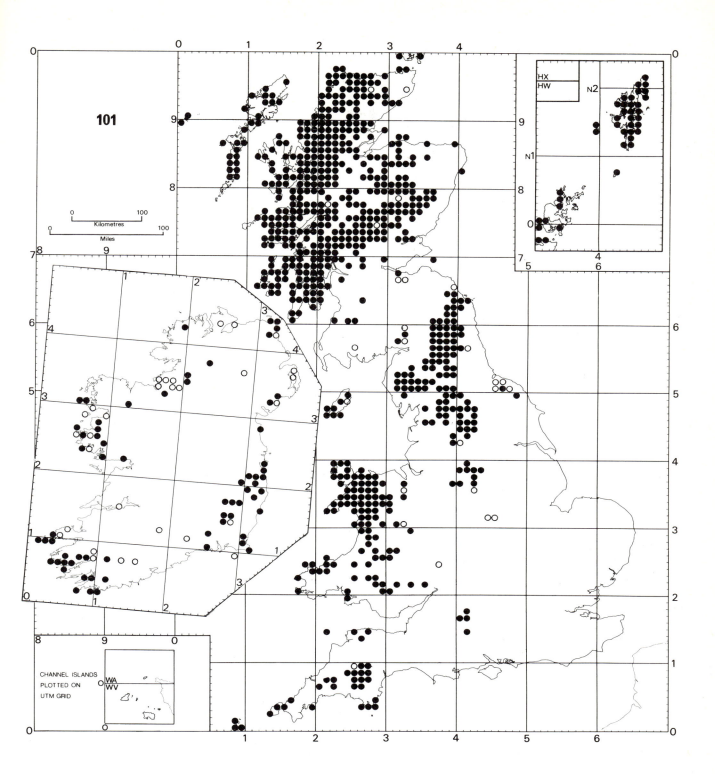

101 *Parmelia omphalodes* (L.) Ach.

Widely distributed in upland areas of the British Isles, occurring mainly on nutrient-poor, coarse-grained siliceous rocks in well-lit situations, but also sometimes over mosses in short turf (in heathland or coastal sites) and on sarsen stones (Wiltshire); occasionally on dry acid bark (especially *Betula*) in a few sites in Scotland and the Lake District. Moderately tolerant of air pollution and still present in the Peak District, although now extinct in Charnwood Forest (Leicestershire). Formerly collected for dyeing in northern England and Scotland. Usually with many other macrolichens and mosses forming the species-rich *Parmelietum omphalodis*, which grades into the *Umbilicarietum cylindricae* on mountains. Some records from northern England may belong to the doubtfully distinct *Parmelia discordans* Nyl. (med. K—, not yellow changing to red). A cool-temperate to arctic species, probably circumpolar in the northern hemisphere; in southern Europe, exclusively high montane.

[Additional Scottish records added in proof: 25/04, 09, 18, 37, 38, 39, 46, 47, 48, 49, 57, 58, 67, 68, 69, 77, 85, 86, 87, 95, 96]

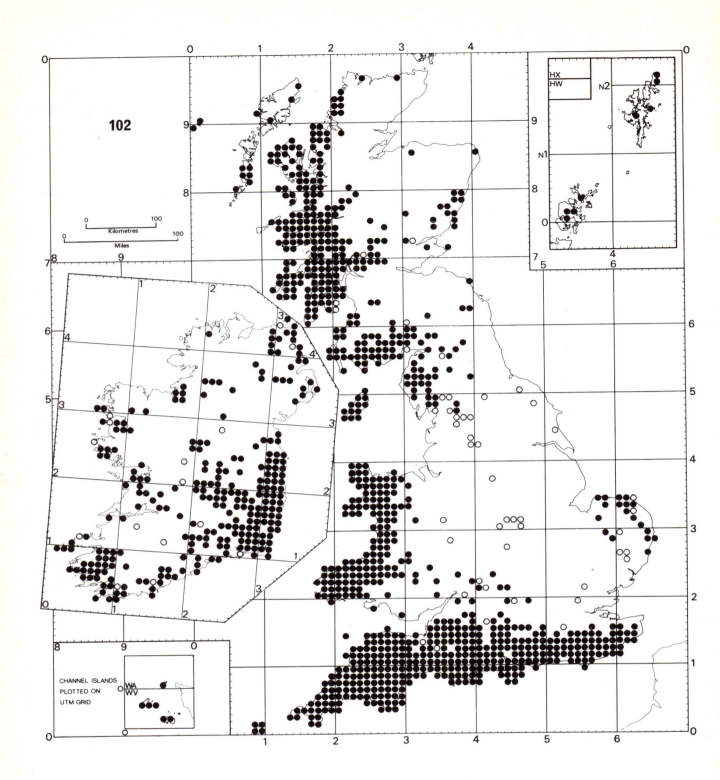

102 *Parmelia perlata* (Huds.) Ach.

Formerly widespread on trees throughout lowland Britain, but sensitive to air pollution and now eliminated where mean sulphur dioxide levels exceed about 35 μg m^{-3}. A characteristic species of the *Parmelietum revolutae,* it has almost identical ecological requirements to *Parmelia caperata* (map 92), being found on sunny coastal rocks and in short coastal turf; also occasionally found in many other epiphytic communities on moderately acid bark. Absent in large unpolluted areas of Scotland and in parts of central Wales and north-east England, where low bark pH leads to the replacement of the *Parmelietum revolutae* by the *Pseudevernietum furfuraceae.* Like *P. caperata,* it is very common in southern Europe, where it occurs even in dry lowland *Quercus ilex* and *Q. suber* woodlands and scrub. A pantemperate species, widespread in Europe where air pollution levels permit, though it is rare and highly oceanic in Scandinavia; known from all continents.

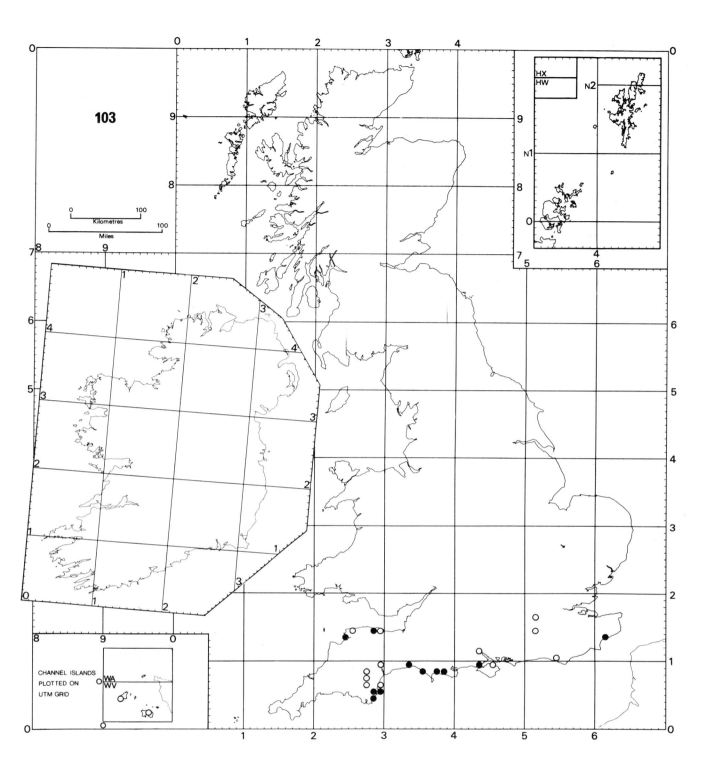

103 *Parmelia quercina* (Willd.) Vainio

A rare, extreme southern species, now extending no further north than Devon in the British Isles, which forms the distinctive *Parmelietum carporrhizantis* community on very well-lit, slightly nutrient-enriched parts of trees, most commonly the upper branches, especially *Acer pseudoplatanus, Fraxinus excelsior* and *Ulmus procera*. An endangered species in the British Isles, now only vigorous and recolonizing in south Devon; relict elsewhere, perhaps due to climatic factors, but also partly to air pollution. A primarily Mediterranean species, locally abundant in southern Europe, but also extending into the Alps in suitable warm situations, and south to the Canary Islands; especially widespread in Asia.

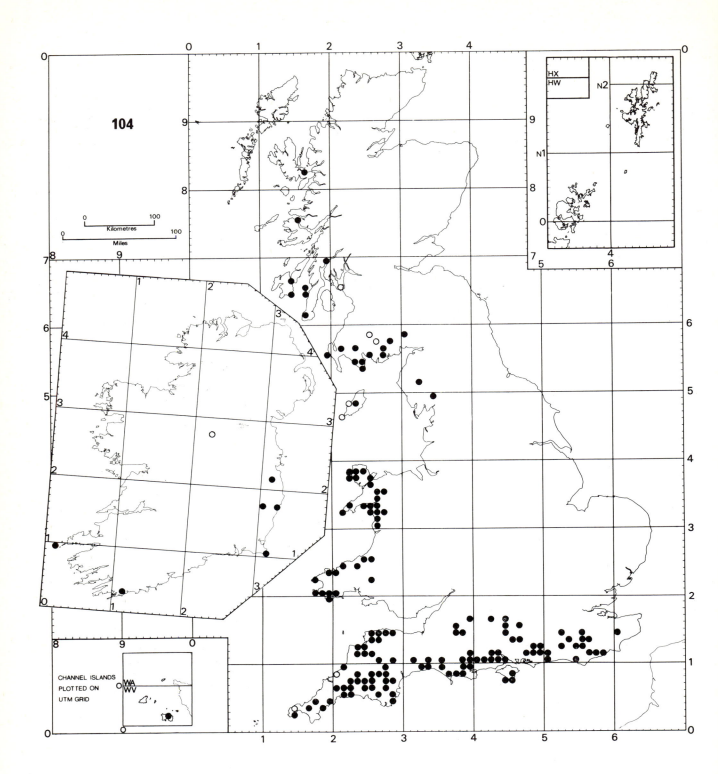

104 *Parmelia reddenda* Stirton

Although now a southern and western species in the British Isles, it was probably formerly widespread in ancient woodland sites throughout lowland Britain, but has been eliminated by a combination of air pollution and woodland clearance. It is essentially a member of the *Lobarion pulmonariae* and an old woodland indicator species, occurring on mossy trunks of broad-leaved trees (rarely mossy rocks) in sheltered, humid, and slightly shaded situations. Much confused with *Parmelia borreri* (map 91) and *P. subrudecta* (map 110) in the past, but differing in its ecology, rather broader blue-green lobes, black under-surface and medulla C— (not red) reaction. A distinctly western species in Europe (absent from the Alps), extending into south Scandinavia, but with a wide distribution, occurring also in east and south Africa, South America and eastern U.S.A.

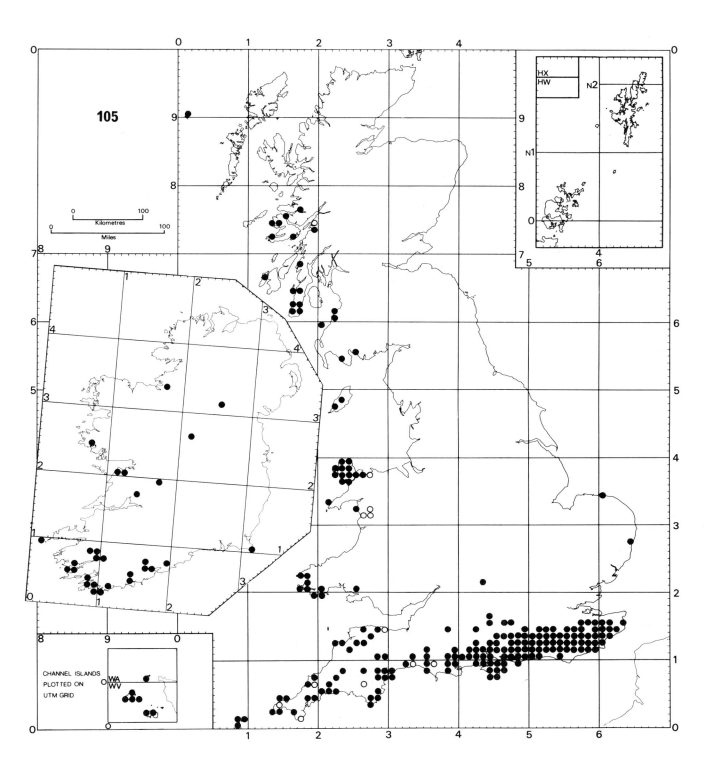

105 *Parmelia reticulata* Taylor

A mainly southern species in the British Isles, most frequent in south-east England, but extending northwards to Argyll and St. Kilda along the west coast in sunny low-rainfall sites. It is almost entirely confined to the epiphytic *Parmelietum revolutae,* but also occurs with the same assemblage of species on sunny coastal rocks in south-west England and Pembroke. Sensitive to air pollution and absent where mean sulphur dioxide levels exceed about 35 μg m⁻³. Easily distinguished from *Parmelia perlata* (map 102) by the minutely areolate upper cortex and chemistry. Essentially an oceanic species in south-west Europe, but absent from Scandinavia; extending east in montane woods to Yugoslavia. Probably pantemperate; also known from Macronesia, south-east Asia, South Africa, South America and Australia, and widespread in southern and eastern North America.

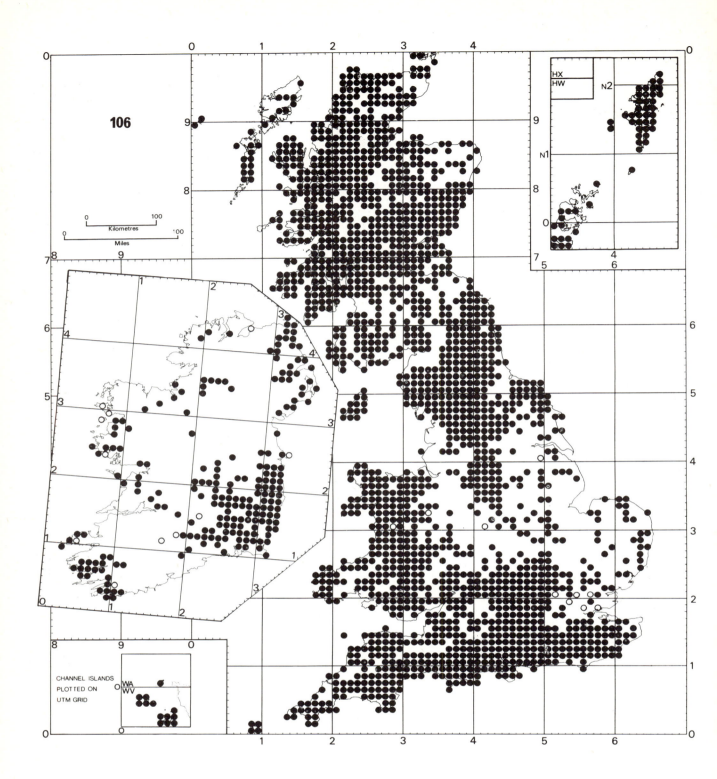

106 *Parmelia saxatilis* (L.) Ach.

A ubiquitous species in the British Isles, absent from trees where mean sulphur dioxide levels exceed *c.* 100 μg m^{-3} but sometimes persisting on sandstones (especially near mortar) at higher levels. Perhaps most common as a component of the *Pseudevernietum furfuraceae,* which occurs on both siliceous rocks and somewhat acidic bark, but also entering numerous other associations, from coastal rocks to the summits of the Scottish Highlands. Less frequent than *Parmelia sulcata* (map 111) in the south and west, but more frequent than it in central and northern England. An arctic to temperate species, probably circumpolar in both hemispheres; much less common in lowland and southern Europe than *P. sulcata,* probably because of the lack of suitable acidic habitats.

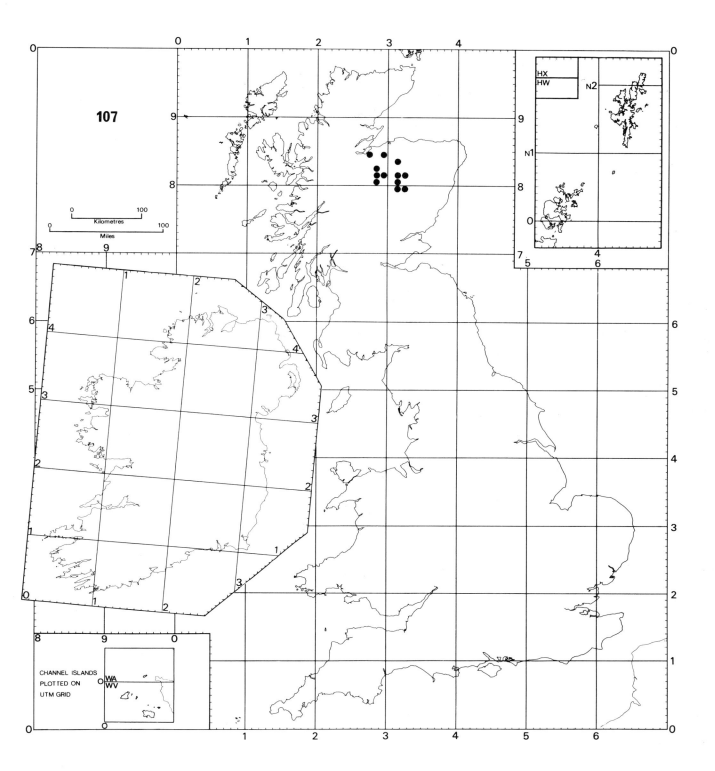

107 *Parmelia septentrionalis* (Lynge) Ahti

An extremely local species in the British Isles, confined to the Scottish Highlands where it occurs almost exclusively on well-lit birch twigs, together with *Cetraria chlorophylla* (map 34) and *C. sepincola* (map 37), in a variant of the *Parmeliopsidetum ambiguae*. Almost always abundantly fertile. Only likely to be endangered by interference with the ancient woodlands in which it occurs or by an increase in air pollution. A circumpolar boreal species confined to the northern hemisphere, with somewhat continental tendencies, and extending south as far as the Alps in Europe. Its preference for *Betula* in Britain is unusual as it is commonest on *Alnus* and *Salix* species elsewhere.

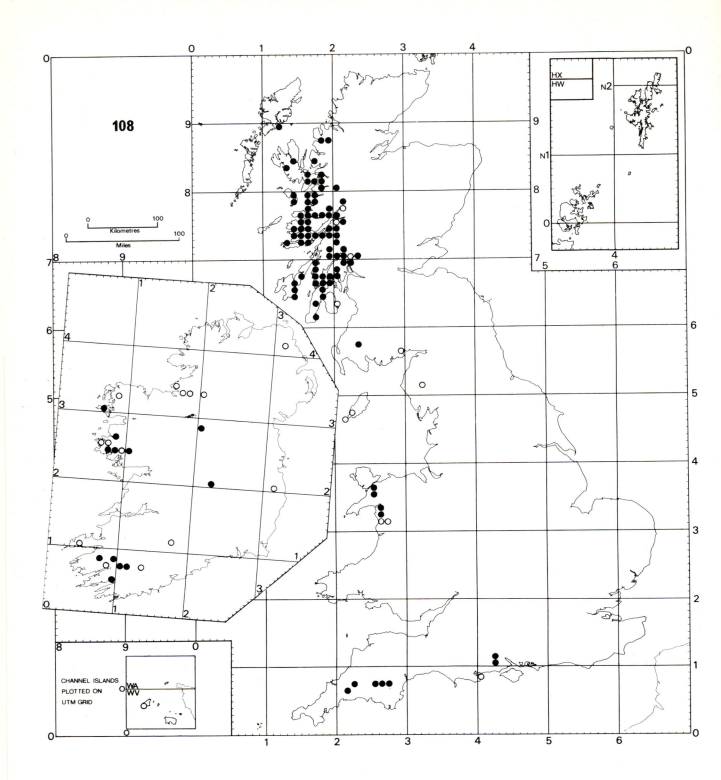

108 *Parmelia sinuosa* (Sm.) Ach.

A strictly western species in the British Isles, frequent only in western Scotland, but extending south to Bodmin, Dartmoor and the New Forest. Confined to more or less continuously humid and moderately exposed cool sites, with over 180 rain days so that bark is leached to about pH 4. Characteristic of twigs of broad-leaved trees supporting the extreme western facies of the *Lobarion pulmonariae* or the *Parmelietum laevigatae* on their trunks; more rarely on mossy rocks. Widespread in Norway and Sweden, in Brittany, and upland humid broad-leaved forests of central Europe. Also known from Central, North and South America, the east African mountains and south-east Asia (Japan to Java), Australia and New Zealand.

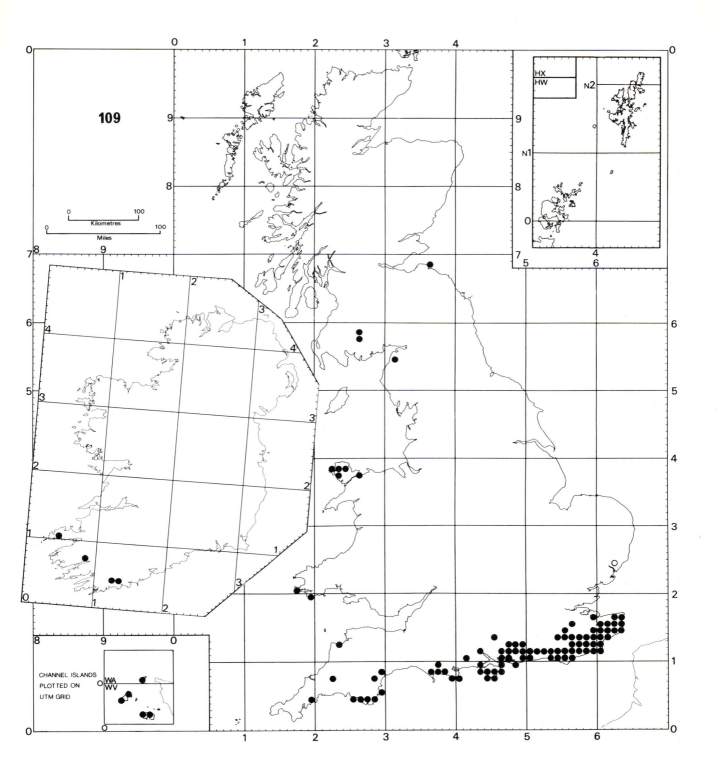

109 *Parmelia soredians* Nyl.

Essentially an extreme southern species, of dry well-lit and moderately acid (pH 5-5.5) broad-leaved trees, on trunks and especially branches; also occurring on wooden fences and smooth-grained siliceous rocks near the coast. Evidently requiring low rainfall, and high summer and mild winter temperatures, its distribution correlating well with the annual mean 4.5 h sunshine day⁻¹ isopleth; extending northwards where these requirements are met to Anglesey, with stations in Cumberland, Dumfries and East Lothian. Primarily a member of the *Parmelietum revolutae*, but also a component of the *P. carporrhizantis* in south-west England. Only likely to be confused with *Parmelia caperata* (map 92), but with narrower, more adpressed and clearly sorediate (K+ red) lobes. Widely distributed through warm and sunny areas of western France into southern Europe; also known from central and south Africa, the Canary Islands, South America and New Zealand.

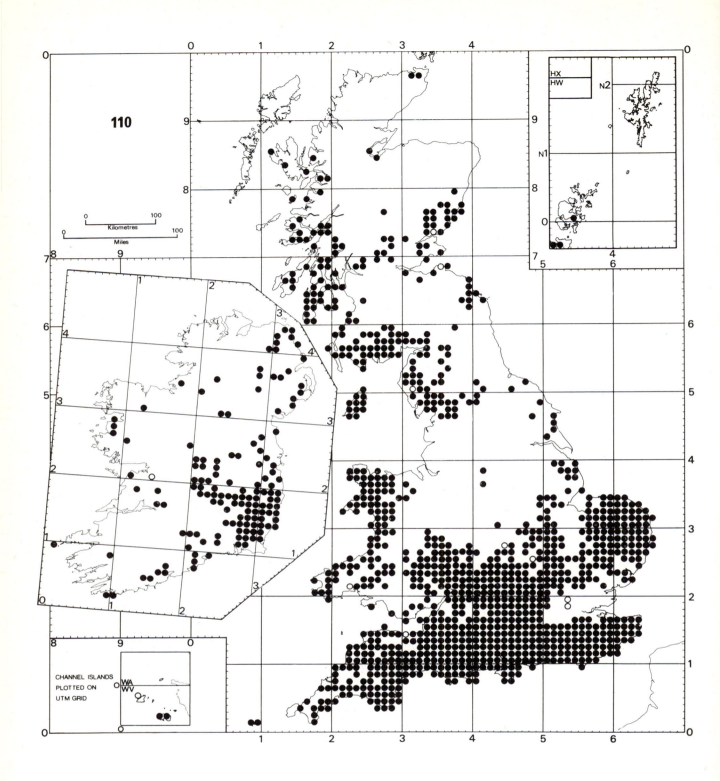

110 *Parmelia subrudecta* Nyl.

A widespread species over most of the British Isles, on well-lit and moderately acid bark of broad-leaved trees, more rarely amongst mosses on siliceous rocks and walls; but absent from large areas of central and northern England where the mean sulphur dioxide levels now exceed about 65 μg m^{-3}. A characteristic species of the *Parmelietum revolutae*. It becomes very rare in Scotland where the generally more acid bark, even of broad-leaved trees, leads to the replacement of this community by the *Pseudevernietum furfuraceae*, but it extends to the Orkneys. Common in Europe, especially in the Mediterranean region, but also in parts of the Alps, with somewhat continental distributional tendencies. Probably circumpolar in the northern hemisphere, but also known in central America, east, north and south Africa, and Australia.

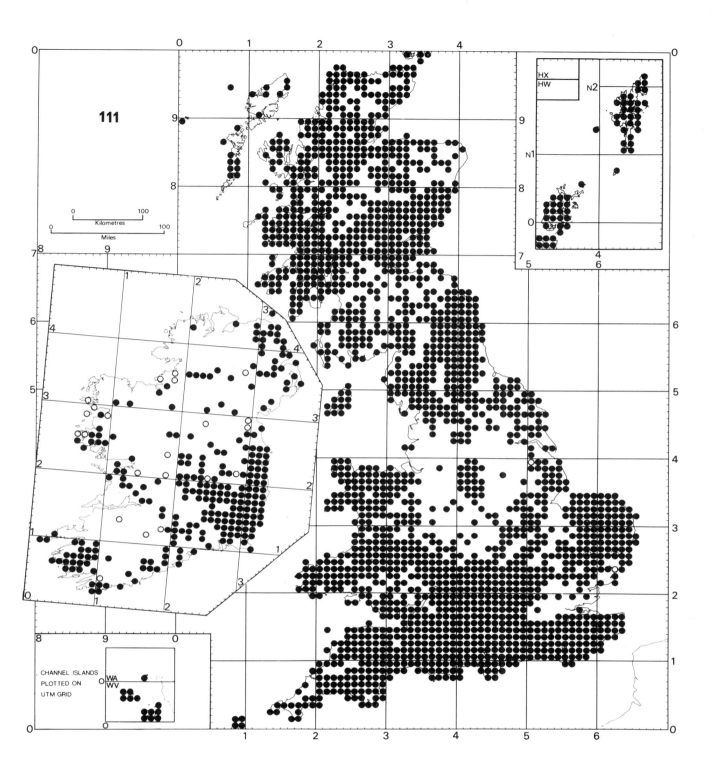

111 *Parmelia sulcata* Taylor

A ubiquitous species throughout the British Isles, except in areas where mean sulphur dioxide levels exceed about 90 μg m⁻³, extending from sunny coastal rocks to *Calluna* stems and rocks on exposed mountain tops. It has a similar ecological amplitude to *Hypogymnia physodes* (map 59), but probably has a slightly higher pH optimum (*c.* 5), and so is an important component in both the *Parmelietum revolutae* and the *Physcieum ascendantis,* though present in many other corticolous and saxicolous communities. An arctic to temperate species, probably circumpolar in both hemispheres; claimed to be the most widely distributed and catholic macrolichen in the world.

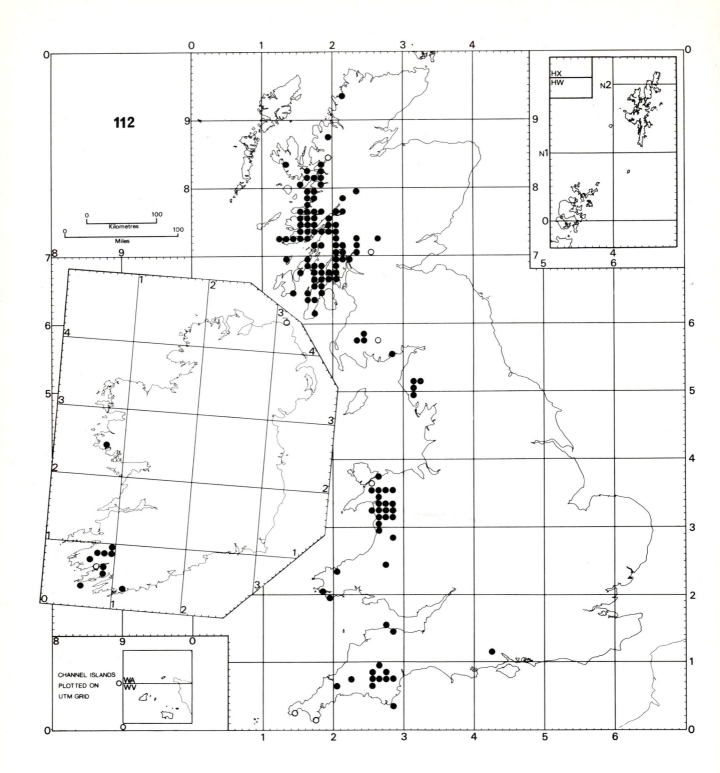

112 *Parmelia taylorensis* Mitchell

A strictly western species, but with an eastern outlier in the New Forest, largely confined to constantly humid, cool and open broad-leaved woods, without extremes of temperature; coinciding well with the 200 rain-days isopleth. It usually occurs on mossy trunks leached to about pH 4, and sometimes on mossy rocks within woodland, or on sheltered coastal rocks in western Scotland. It is a characteristic species of the strongly oceanic *Parmelietum laevigatae* and only likely to be threatened in the short-term by increasing air pollution levels. Apparently almost confined to Europe, the species extends from Madeira through north Spain to west France and Sutherland, Scotland (its most northerly station), but is also present in cool humid forests of the northern Alps.

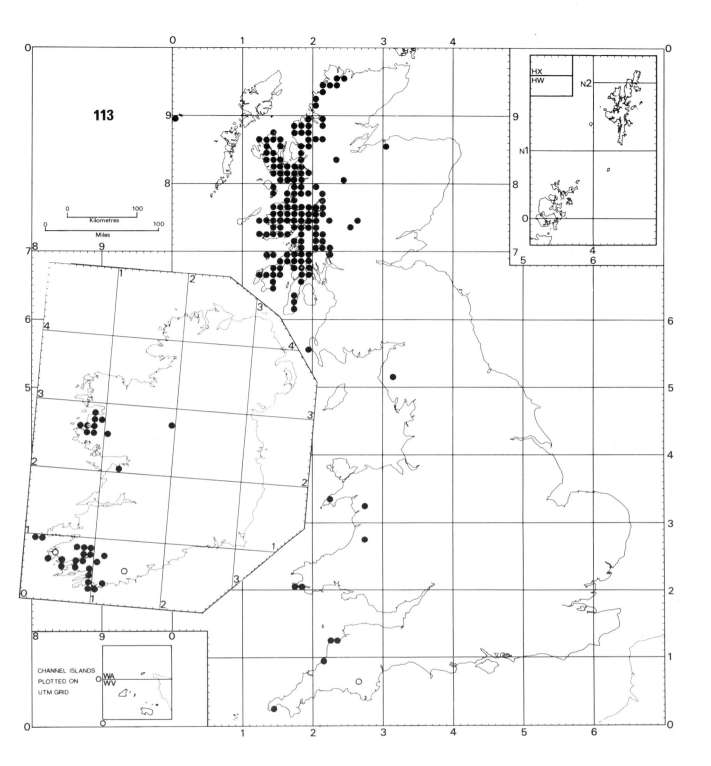

113 *Parmeliella atlantica* Degel.

A more strongly western species than the allied *P. plumbea* (map 114), originally described from south-west Ireland. Locally abundant in western Scotland and Ireland on moderately shaded, mossy broad-leaved trees in cool, sheltered and continuously humid woodlands, rarely on mossy rocks, extending south to north-west Devon and Cornwall. A member of the *Lobarion pulmonariae*, its status is likely to be threatened by any increase in air pollution and the clearance of ancient woodland sites. Largely confined to extreme oceanic areas in Europe (with western Scotland as its major stronghold), extending to Norway, south to north Africa and Macronesia, and east to Greece, but confined to montane forests in the south of its range.

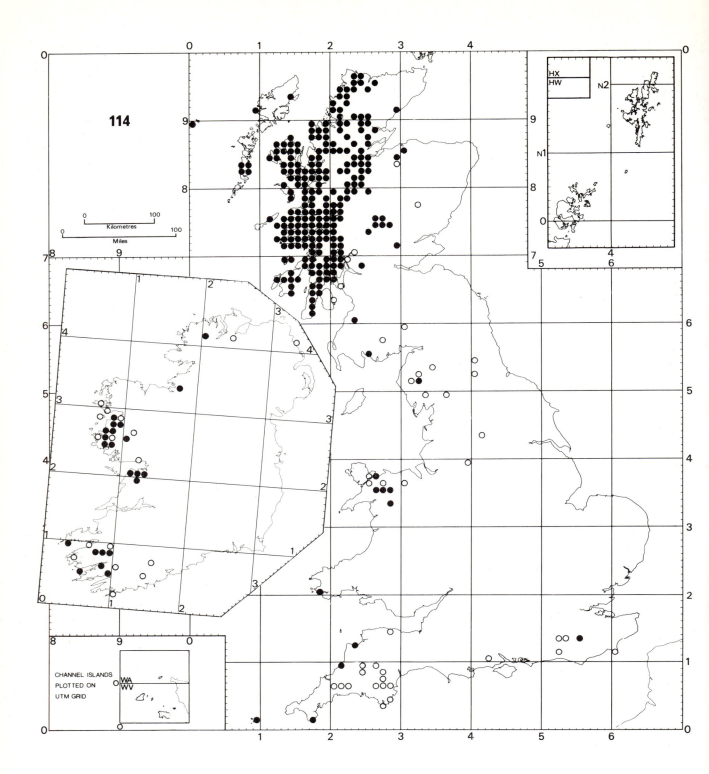

114 *Parmeliella plumbea* (Lightf.) Vainio

A southern and western species, locally abundant in north-west Scotland and western Ireland, which requires the maintenance of a mild climate and high humidity throughout the year. A characteristic component of the *Lobarion pulmonariae,* it occurs mainly on mossy broad-leaved trees and granitic or schistose rocks in moderately open woodland or parkland at sites with a long history of ecological continuity; also on maritime rocks in extreme oceanic areas. Its decline in north, south and south-west England is due to a combination of the effects of air pollution and the clearance of ancient woodland sites. Widespread in oceanic and southern montane forested areas of Europe, extending from northern Norway south to north Africa, and Macronesia; otherwise known only in a few sites on the east coast of North America.

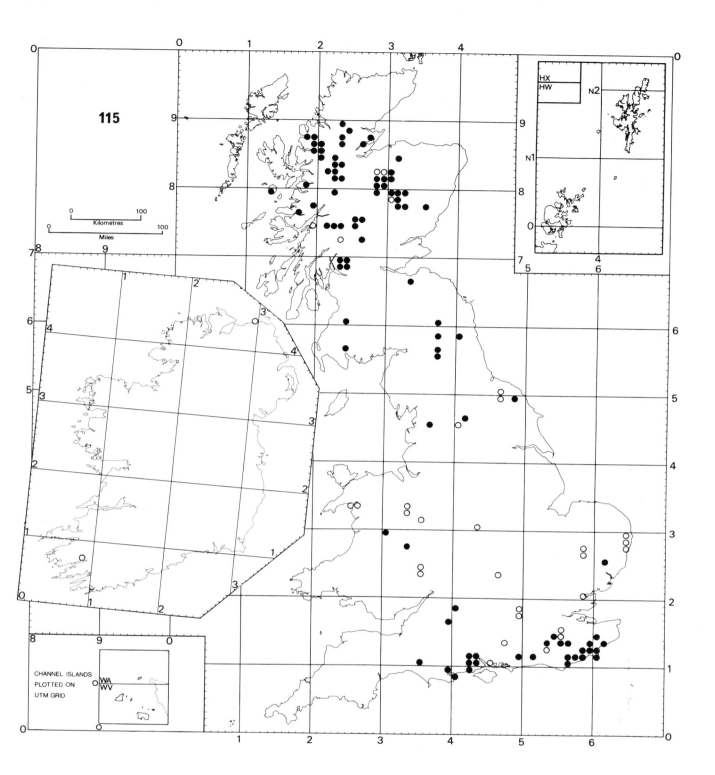

115 *Parmeliopsis aleurites* (Ach.) Nyl.

A species of very acid bark and decorticate wood, whose ancient habitat in Britain seems to have been in the *Parmeliopsidetum ambiguae* on *Pinus* and *Juniperus* trunks and boughs, and decorticate stems, in the Caledonian Forest of the Scottish Highlands; it is still frequent in the relics of this forest. It is also rare on natural decorticate wood elsewhere, as in the New Forest, but its habitat otherwise in southern England is almost entirely on artificial substrates, such as oak and conifer wood fence rails. It may have undergone some slight extension in range as a result of increased bark acidification in south-east England, but has declined in southern-central England and East Anglia, probably due to air pollution and a reduction in the use of wooden fencing. It is very rare in Ireland, where there are no recent records. A species of montane and boreal coniferous woods throughout the northern hemisphere, circumboreal, and widespread in less polluted parts of Europe.

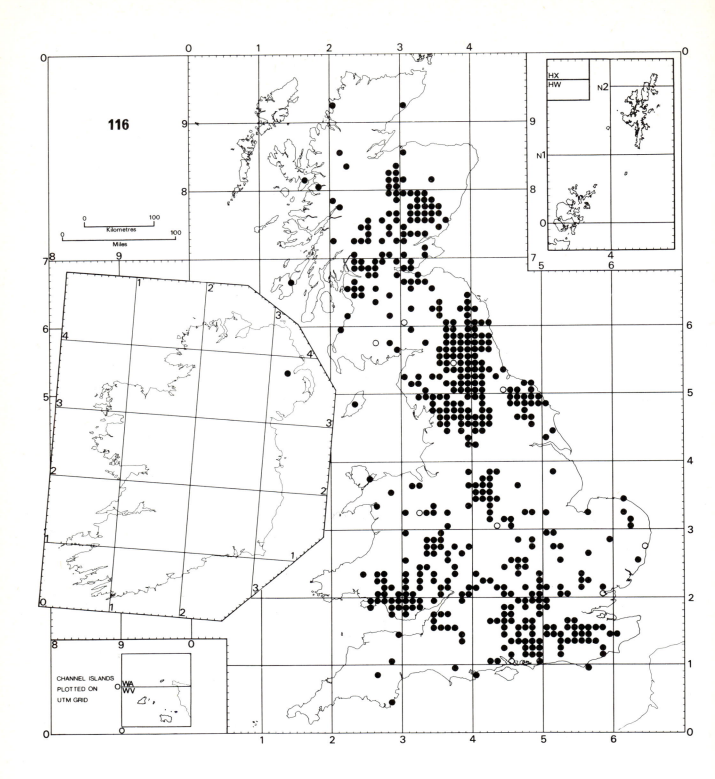

116 *Parmeliopsis ambigua* (Wulfen) Nyl.

Formerly only a species of the *Parmeliopsidetum ambigua,* a community of decorticate conifer wood in central and eastern Scotland. During the present century it has undergone a remarkable expansion onto deciduous trees in moderately polluted areas (with mean winter sulphur dioxide levels *c.* 60-70 μg m^{-3}), presumably a response to increased bark acidification (to pH 3.0-4.0), and can even be found on siliceous rocks. Now widespread in such areas in northern, central and south-east England where it occurs in a special facies of the *Pseudevernietum furfuraceae.* Exceptionally rare in regions of England with relatively clean air (i.e. the south-west and west) where it again mainly occurs on decorticate wood; perhaps introduced by man into these areas and Northern Ireland. Circumboreal in the northern hemisphere in the coniferous zone. Widespread in Europe, extending southwards into the southern Alps.

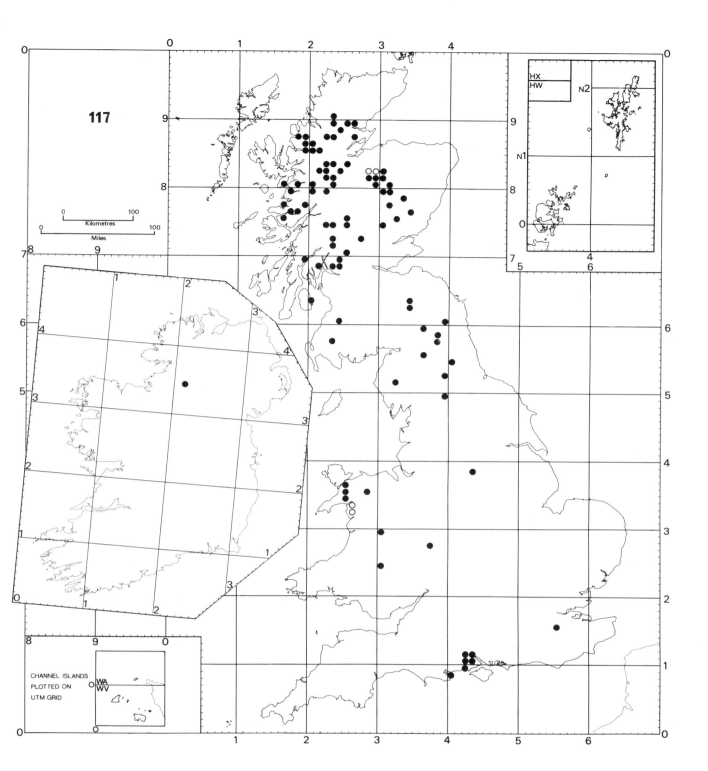

117 *Parmeliopsis hyperopta* (Ach.) Arnold

This species is almost entirely confined to decorticate coniferous wood, and is only very rarely found on other substrates such as decorticate wood of *Quercus* and on the bark of *Betula, Fagus* and conifers. It is a member of a species-poor British variant of the Continental-subboreal *Parmeliopsidetum ambiguae*. It is a widespread, but local, species in the Scottish Highlands and northern England, and is most frequent in upland, poorly maintained, old conifer plantations containing dead wood and tree stumps. In southern England it is not uncommon on palings in the New Forest. Unlike *P. ambigua* (map 116), *P. hyperopta* has not increased its range markedly this century. It is a widespread circumboreal species in the northern hemisphere.

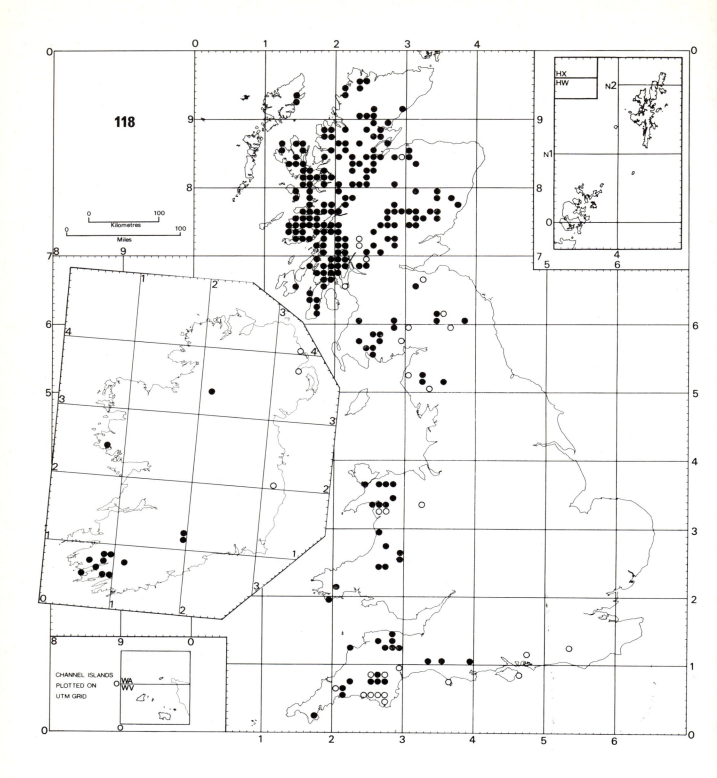

118 *Peltigera collina* (Ach.) Schrader

A corticolous, locally frequent, old woodland indicator species on *Quercus, Fraxinus,* etc. It is a frequent member of the *Lobarion (Nephrometum lusitanicae),* and occasionally is locally muscicolous over non-calcareous rocks. Rarely found in the fertile state. Common in Scotland (except the central lowlands and north-east), occasional in western England and Wales, but absent now from the region east of Dorset and the Lake District. Scattered and locally common in western Europe, in areas of low air pollution only. It is not very oceanic, but in drier southern areas is confined to montane forests.

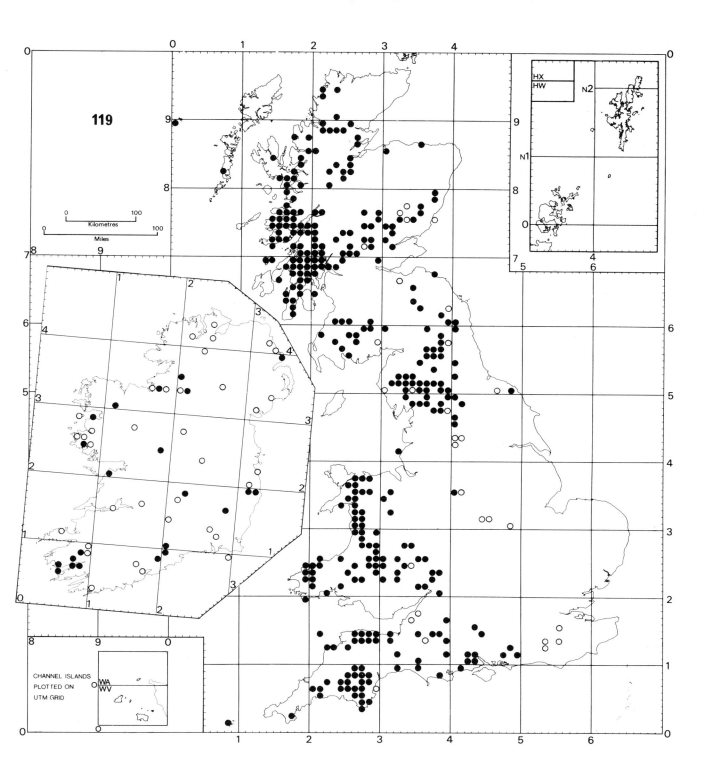

119 *Peltigera horizontalis* (Huds.) Baumg.

Corticolous, saxicolous and terricolous on sheltered moist banks among bryophytes; often in large patches on lower parts of boles or on roots of trees, especially old *Fraxinus, Ulmus, Quercus* and *Fagus*. Frequent to locally common, except in central, eastern and south-eastern England where it is now rare or absent. It extends further east than *P. collina* (map 118), into West Sussex, Hampshire, Gloucestershire and north Staffordshire. An old woodland indicator species often forming part of the *Lobarion* community, especially on damper or more basic substrates. It is usually fertile. It has decreased in eastern England, and is absent from areas where sulphur dioxide pollution exceeds 40 μg m^{-3} (mean winter levels), except on limestone substrates. Widespread in western Europe.

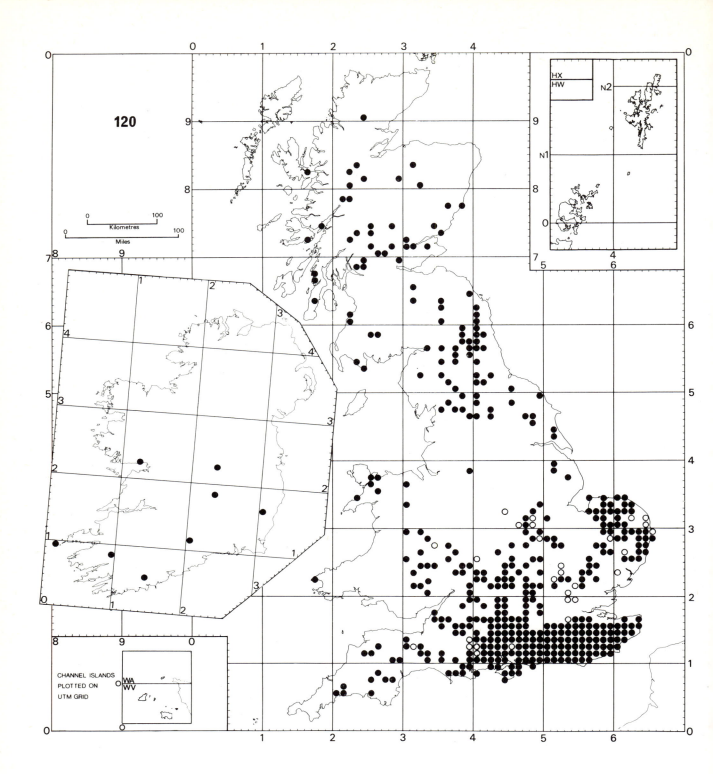

120 *Pertusaria coccodes* (Ach.) Nyl.

A species that occurs predominantly on smooth or rough bark of broad-leaved trees, chiefly *Acer pseudoplatanus, Fraxinus, Quercus,* and *Ulmus glabra* on waysides and in parkland. Although preferring well-lit sunny sites, it is also tolerant of some shade and may occur in old woodlands. Occasionally the species may grow on lignum, as well as on soft slate and sandstone gravestones. It is only rarely fertile. It is a member of the *Parmelietum revolutae* and the facies of this association which merges into the *Physcietum ascendentis. P. coccodes* is moderately pollution-tolerant, surviving concentrations of sulphur dioxide of up to 60 µg m⁻³ and moderate contamination by inorganic fertilizers. It is endemic to Europe, where it is widespread from south Scandinavia to the Mediterranean.

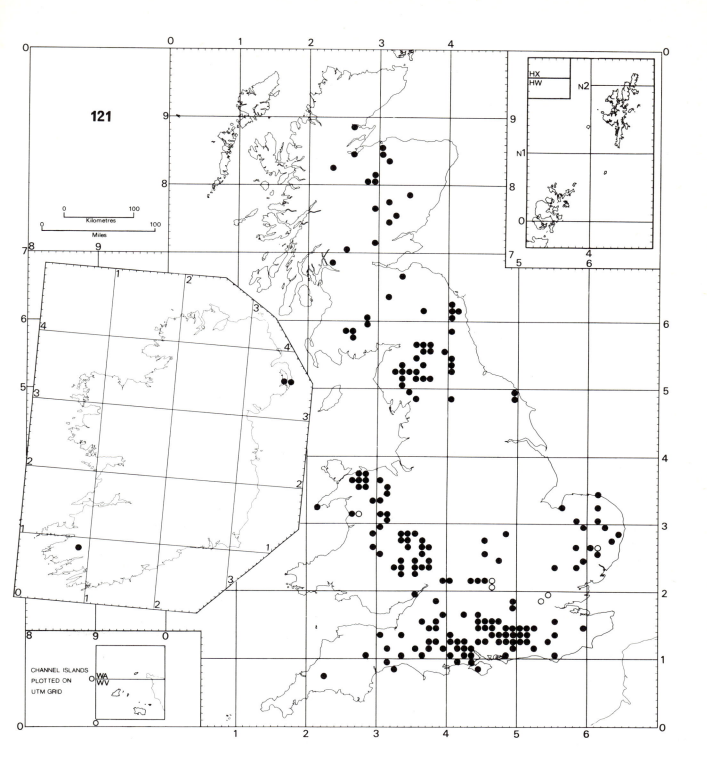

121 *Pertusaria flavida* (DC.) Laundon

A species of smooth or rough, usually very well-illuminated, bark (pH *c.* 4.5-6.0) of mature trees such as *Quercus, Fagus* and *Fraxinus,* in open woodlands and parklands. It is tolerant of sulphur dioxide levels below about 45 µg m⁻³, and though found in old forests, is by no means restricted to them. It is very rarely fertile. It has a rather eastern tendency in Britain, becoming rare or absent towards the west coast, and is very rare in Ireland. This agrees with its broadly continental type of distribution in Europe, where it is very widespread within the deciduous forest zones, but becomes exclusively montane southwards, occurring there on *Abies alba* as well as on broad-leaved trees. Probably restricted to Europe and the Canary Islands; incorrectly reported from North America.

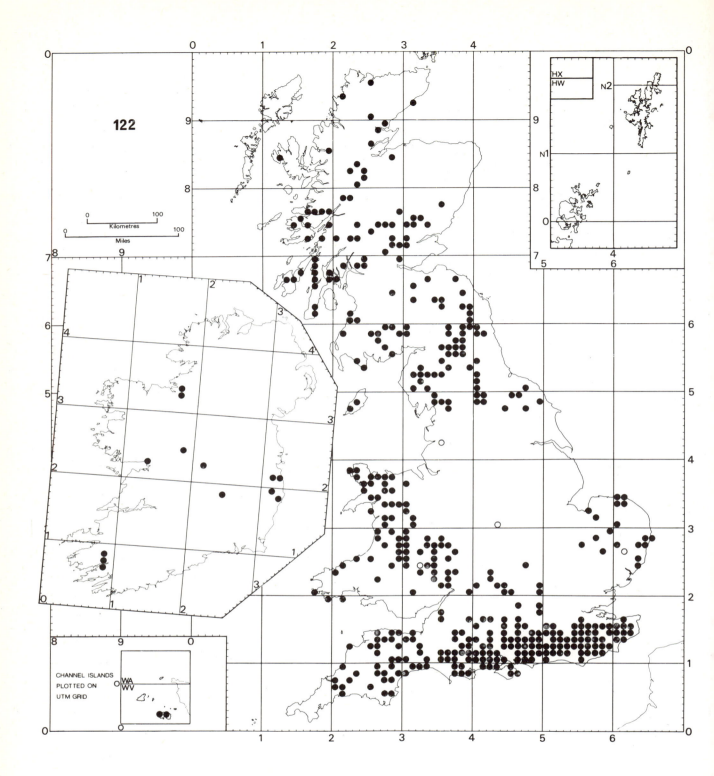

122 *Pertusaria hemisphaerica* (Flörke) Erichsen

A species of rough, usually well-illuminated bark on the south and west sides of the trunks of mature to old deciduous trees, such as *Quercus* and *Fraxinus,* with moderately acid (pH 4.5-6.0) bark. It is a member of the *Parmelietum revolutae,* commonest in open woodland, but also occurs on less shaded boles in denser forests and on trees in open parkland or even by roadsides and in meadows. It is moderately pollution-tolerant, withstanding mean winter sulphur dioxide levels below about 45-50 μg m^{-3}, and tolerant of dry conditions, but not of hypertrophication by fertilizers. It can occur in the nutrient-requiring *Physcietum ascendentis* association, but is not particularly common in it. Still very common in less polluted areas of Britain, especially in southern England, but becoming rarer in northern Scotland and absent from the hyperatlantic, more northern Hebridean islands, and from exposed situations on the west coast of the mainland. It is widespread in Ireland, but much under-recorded there. Many old reports of *Pertusaria velata* (map 123) in the British Isles belong to this species. In Europe, it is a common and widespread species of the deciduous woodland zone from Norway southwards, except in the more polluted regions of the northern European plain; in the south it is apparently confined to more montane forests where it occurs both on broad-leaved trees and on *Abies alba.* Probably endemic to Europe, incorrectly reported from North America.

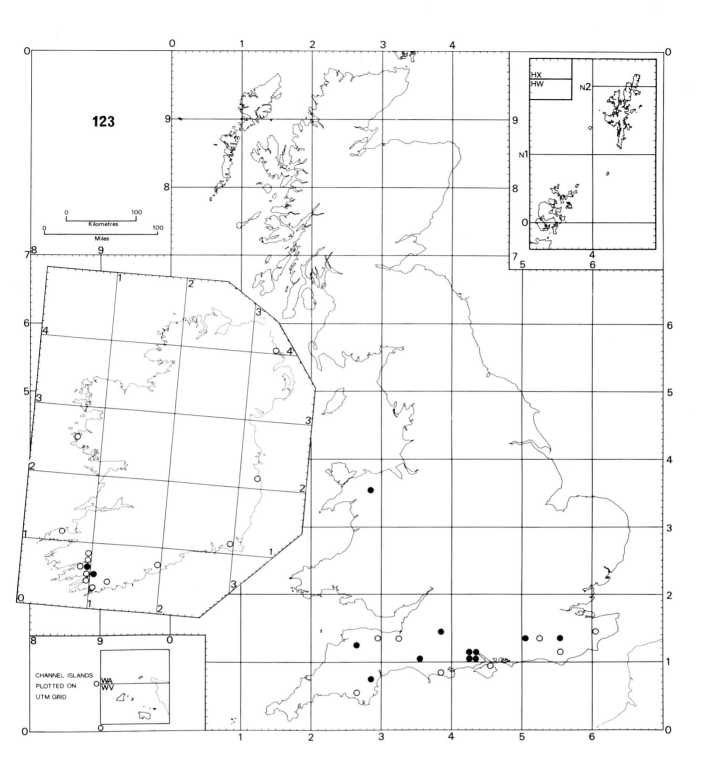

123 *Pertusaria velata* (Turner) Nyl.

A very rare, extremely southern, corticolous species of ancient woodlands and parklands, on sheltered, but well-illuminated, boles of mature trees with weakly acid barks (pH 5-6), such as *Fagus, Quercus* and *Fraxinus*. It is normally associated with other *Pertusaria* and *Graphis* species, *Pachyphiale cornea, Thelotrema lepadinum* (map 162), etc. in a damp, often bryophyte-rich variant of the *Graphidion scriptae* merging with elements of the *Lobarion pulmonariae*. It is apparently always fertile—indeed it could hardly be detected otherwise—but has been confused with *P. hemisphaerica* in the past. It is mainly found in the New Forest, where it is locally frequent, especially on mature *Fagus*, in at least thirteen of the ancient woodland areas; otherwise it is very rare and scattered in Devon, Dorset, Wiltshire and Sussex, with a northern outlier in the Fairy Glen in Denbigh. It is very rare in Ireland, but locally frequent in the old oakwoods near Killarney. In Europe it is very rare and oceanic. It is correctly recorded for one site (beech forest) in south-west Sweden, where it has not been observed recently, and in Germany only in Schleswig-Holstein. The only other area where it is currently known is in Brittany, where it occurs in several old forests. Many older records from western Europe (e.g. Normandy, Maine) are in fact *P. hemisphaerica*. It has undoubtedly decreased in western Europe due to changes in forest management, and also to air pollution (e.g. in northern Germany) to which it is probably very sensitive. Very rare in southern Europe, but reported from Portugal. Apparently with a wide tropical-subtropical oceanic distribution; known in North and South America, Australasia, Macronesia, Africa and Asia.

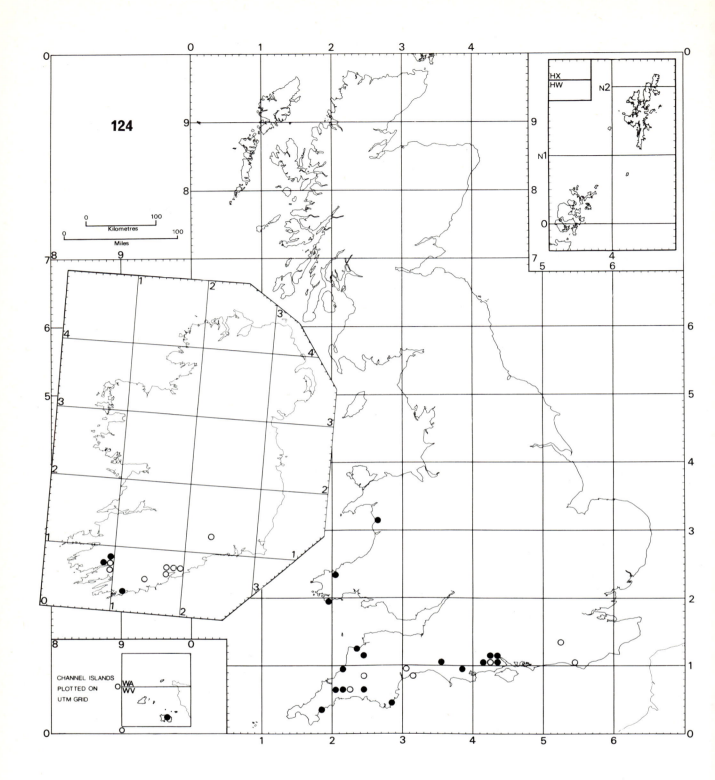

124 *Phaeographis lyellii* (Sm.) Zahlbr.

A rare, local corticolous species on smooth bark of *Fagus, Corylus,* and young *Quercus* and *Castanea* in old woodland areas. A rare component of the *Graphidetum scriptae.* Mainly confined to southernmost England, especially the south-west, but extending eastwards to the New Forest; also in south-west Ireland. Now extinct in south-east England, and under further threat from woodland clearance and air pollution at other sites. Only known elsewhere from western France (especially Brittany) and Portugal.

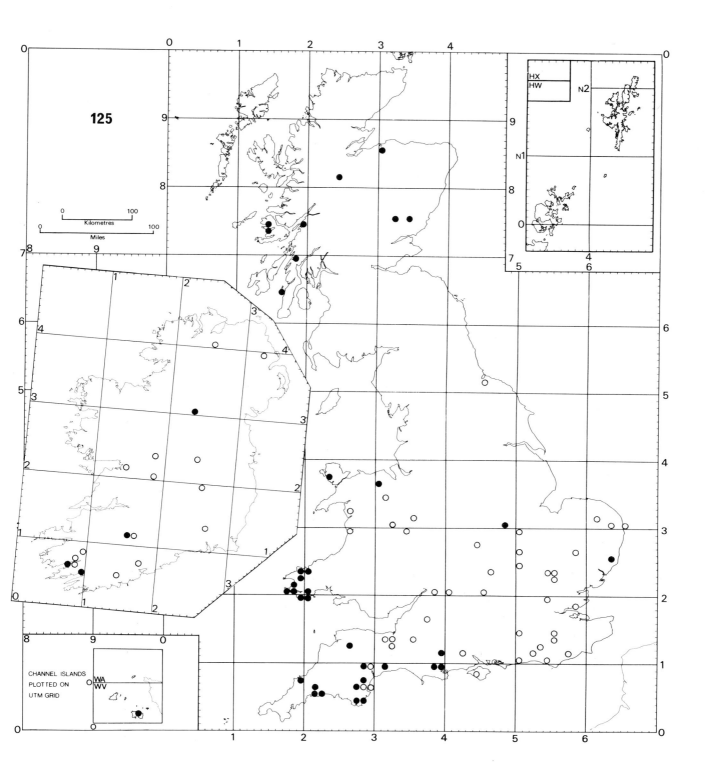

25 *Phlyctis agelaea* (Ach.) Flotow

A species of smooth bark (particularly *Corylus, Fraxinus* and *Salix*) which is weakly acid (pH 4.5-6), in open lowland woodland or even scrub areas, especially in marshes or by streams, occurring in either the *Parmelietum revolutae,* the *Lecanoretum subfuscae* or the *Graphidion scriptae.* It is always fertile, a feature which helps to distinguish it macroscopically from the usually sterile, and always sorediate *P. argena.* It was formerly widespread in the British Isles, and probably common (though old records are not numerous enough to give a clear idea of its former frequency), but it has become very rare and confined to south-west England, western Wales, and scattered sites in Scotland; similarly widespread (at least formerly) in Ireland, it has only been recorded recently near Killarney. In the Suffolk site it has not been seen for 12 years. Its decline is hard to explain in the British Isles, as it is still a quite common species of boles and small branches of broad-leaved trees throughout the less-polluted areas of western and central Europe, from south Scandinavia (including Denmark), throughout France and to Italy at least, and seems there to be no more pollution-sensitive than such species as *Pertusaria hemisphaerica* and *Parmelia revoluta,* although it cannot now be found in the New Forest where species far more sensitive to pollution still flourish. It is suboceanic to continental in distribution, rather than oceanic, but in southern Europe seems to be confined to damper, more montane woodlands. Distribution elsewhere uncertain, but reported at least from North America.

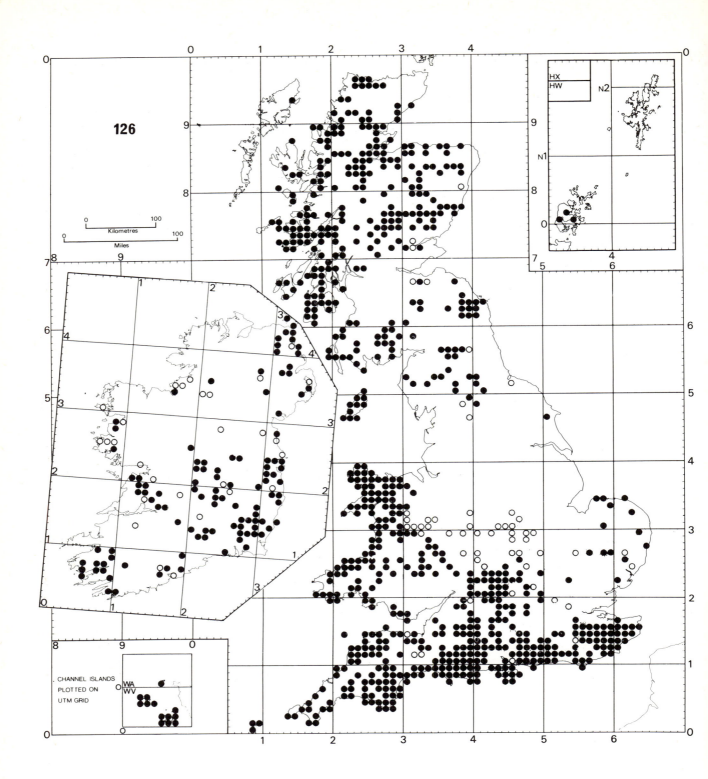

126 *Physcia aipolia* (Ehrh. ex Humb.) Fürnrohr

A widespread species often on isolated, mature, well-lit and nutrient-enriched or hypertrophicated trees in areas not exceeding pollution levels of 45 µg m⁻³ of sulphur dioxide. It grows on *Ulmus, Acer pseudoplatanus, Sambucus nigra, Fraxinus* and orchard trees, and more rarely on *Quercus, Crataegus* and *Fagus*. Although occurring on boles, it is much more often a colonizer of twigs and small branches. It frequently occurs in the *Physcietum ascendentis,* particularly on *Sambucus nigra,* and occasionally in the *Teloschistetum flavicantis.* The species is widespread in most of Britain, but its absence or scarcity in central and eastern England and the Scottish Lowlands reflects its sensitivity to air pollution; inorganic fertilizers, the ravages of Dutch Elm Disease and removal of hedgerows may further reduce its range in these areas. The boreal European var. *alnophila,* with smaller spores and apothecia occurring to the tips of the lobes, has not yet been found in Britain. Widespread throughout Europe and also present in north Africa; circumboreal in the northern hemisphere and common in Central and North America.

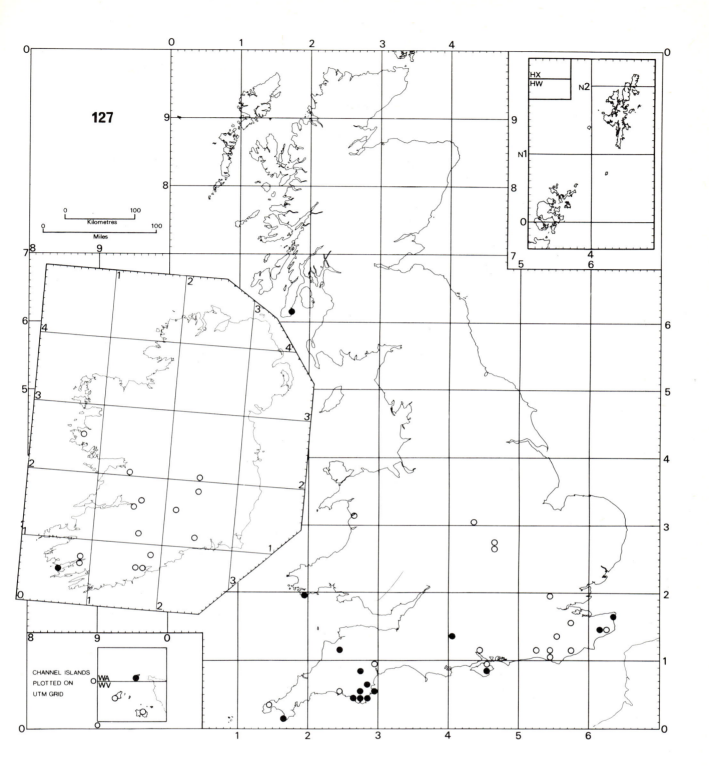

127 *Physcia clementei* (Sm.) Maas Geest.

Formerly widespread in southern England and Ireland, but now extremely rare and decreasing. Primarily a member of the *Teloschistetum flavicantis,* characteristic of the sunniest and driest parts of England, it occurs on well-lit and well-ventilated nutrient-rich bark; there are also four records on soft basic sandstones and walls (South Devon and Kintyre). Old apple orchards and dust-enriched roadside trees were once major habitats. It occurs mainly on *Acer pseudoplatanus, Malus, Quercus ilex, Sambucus nigra* and *Ulmus.* Today its stronghold is in South Devon, where it is very local, with outlying localities in Pembroke and Kintyre. Increasing orchard hygiene, especially the use of pesticides, appears to have been a major factor in its decline, but it is also probably very sensitive to sulphur dioxide air pollution. A Mediterranean-Atlantic species, extending from Italy, Portugal, Spain and Sardinia north into the southern Alps; formerly also into Belgium, north-west Germany and the Netherlands, but now scarce to extinct in these areas; absent from Scandinavia. Also on *Juglans* in California, but perhaps introduced there from Europe.

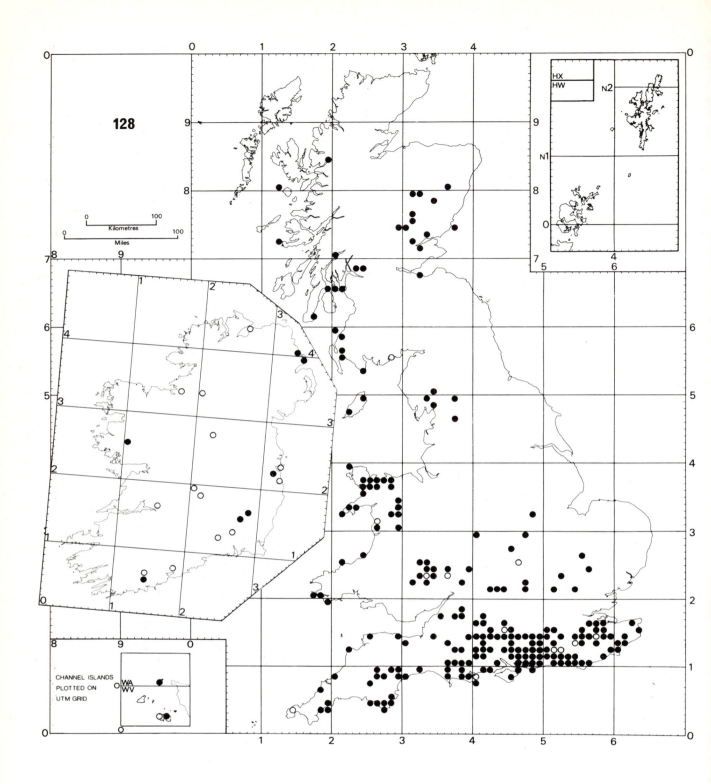

128 *Physcia tribacia* (Ach.) Nyl.

A local species often on isolated, mature, well-lit and nutrient-enriched trees, occasionally on rocks in areas with little air pollution. Trees favoured are *Ulmus, Acer pseudoplatanus, Sambucus nigra* and occasionally *Quercus*. It commonly occurs in the *Physcietum ascendentis* and occasionally in the *Teloschistetum flavicantis*. On rocks, it enters the *Physcietum ascendentis* of nutrient-enriched calcareous rocks and walls, and the *Candelarielletum corallizae* on ornithocoprophilous rocks. Most frequently recorded in south-east England, but unfortunately the recent large-scale loss of *Ulmus* trees due to Dutch Elm Disease may have drastically reduced the range of this species. In coastal sites, such as Cornwall, South Devon, Pembroke and western Scotland, it is more frequent than elsewhere on old barn walls and bird-perching rocks. It is probably widely distributed in Ireland. Mainly a south European and Mediterranean species, but extending northwards into southern Germany, Hungary, Switzerland and the Netherlands, but now rare in many of these areas; absent from Scandinavia. Recently found in Japan; reports from arctic areas (from Canada east to Siberia) require confirmation.

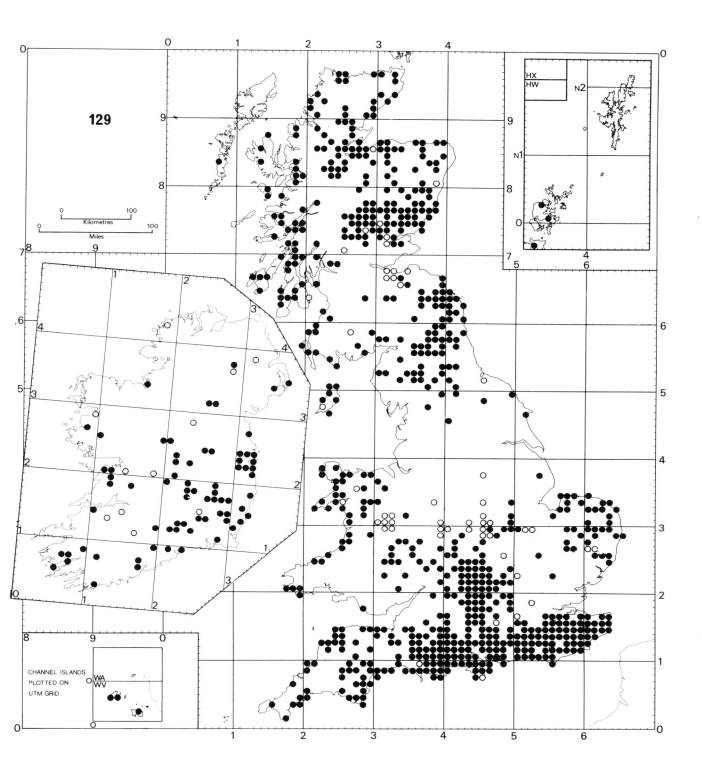

129 *Physconia pulverulacea* Moberg

A species of isolated, mature or aged, well-lit and nutrient-enriched or hypertrophicated trees; *Ulmus, Acer pseudo-platanus, Sambucus nigra, Fraxinus* and orchard trees are most favoured, but it also colonises wound scars of *Fagus* and *Quercus*. It is a bole and branch species, entering the *Physcietum ascendentis*. When saxicolous, it occurs on nutrient-enriched limestone, acid or basic bird-perching stones, tops of concrete posts and basic gravestones, and other sites enriched by bird droppings and organic manures. As a saxicolous species, it forms an element of the *Physcietum caesiae*. The species is widespread in Britain in areas where sulphur dioxide pollution does not exceed 50 µg m^{-3}, and where it is not affected by inorganic fertilizer. It is absent from central England and south Wales, and is less common than *Physcia aipolia* (map 126) in western areas of the British Isles. It is widespread and probably frequent in Ireland. Well distributed throughout Europe; in the south it is often found as a forest species, even in *Lobarion* communities.

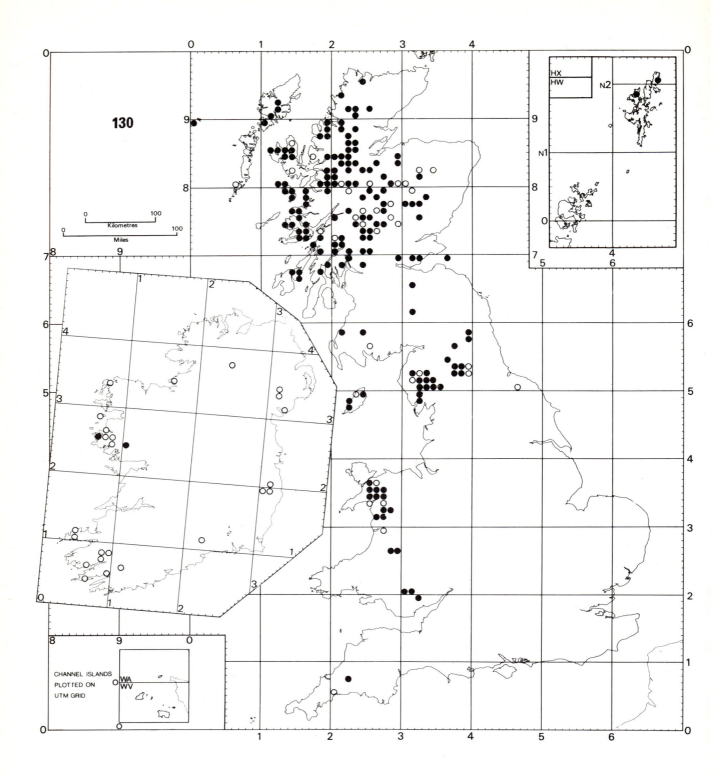

130 *Placopsis gelida* (L.) Lindsay

A species of well-lit, damp acid rocks and boulders, often in or near streams, in areas of late snow lie or on low rocks in damp moorland. It enters a moist facies of the *Lecideetum lithophilae* and appears to have a preference for rocks somewhat rich in heavy metals, hence its occurrence in the *Acarosporetum sinopicae*. Its altitude ranges from the seashore to *c.* 1200 m. In Britain it is widespread, but seldom frequent, in western and central Scotland; elsewhere it is local on mineral-rich outcrops in the Lake District and North Wales. Widely distributed, occurring in arctic and temperate oceanic Europe, North America, east Asia, Macronesia, New Zealand, Tasmania and Chile.

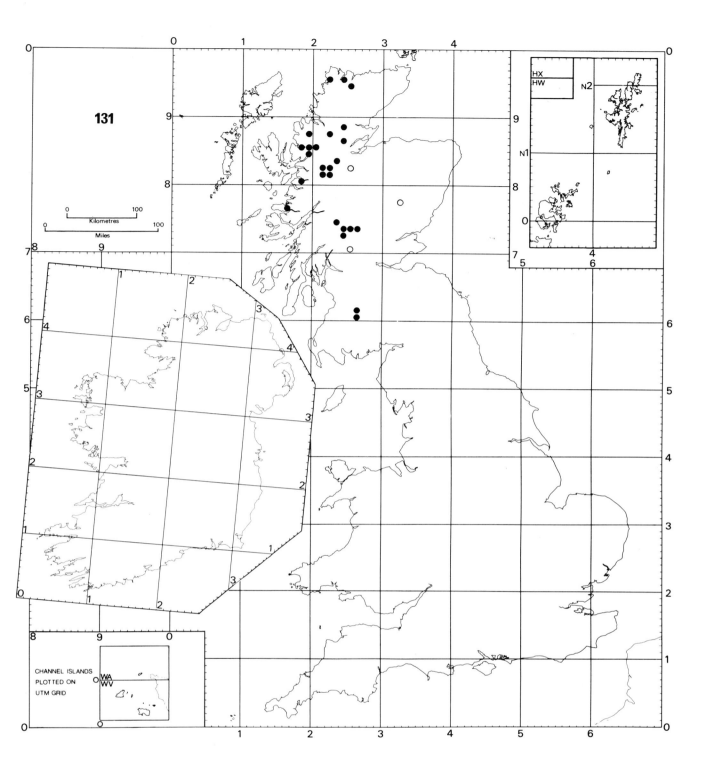

131 *Platismatia norvegica* (Lynge) Culb. & C. Culb.

A boreal oceanic species in the British Isles known only from Scotland, where it is very rare. It occurs in moss-lichen heaths of the *Arctoeto-Callunetum,* but can also be found on exposed hard siliceous rocks in communities of the *Umbilicarion cylindricae,* and on well-lit leached low pH bark, especially of *Pinus* and *Betula* in the relict old Caledonian forest. The species was overlooked by many early workers and so may once have been more widespread. The abundant development of isidia and the generally paler grey thallus readily separates the species from the common *P. glauca* (L.) Culb. & C. Culb. Through most of its range the species is characteristic of humid, oceanic spruce forests without extremes of temperature. It is incompletely circumboreal, occurring only in western Norway, Sweden, Finland, Newfoundland, and on the west coast of North America from southern Alaska to Oregon.

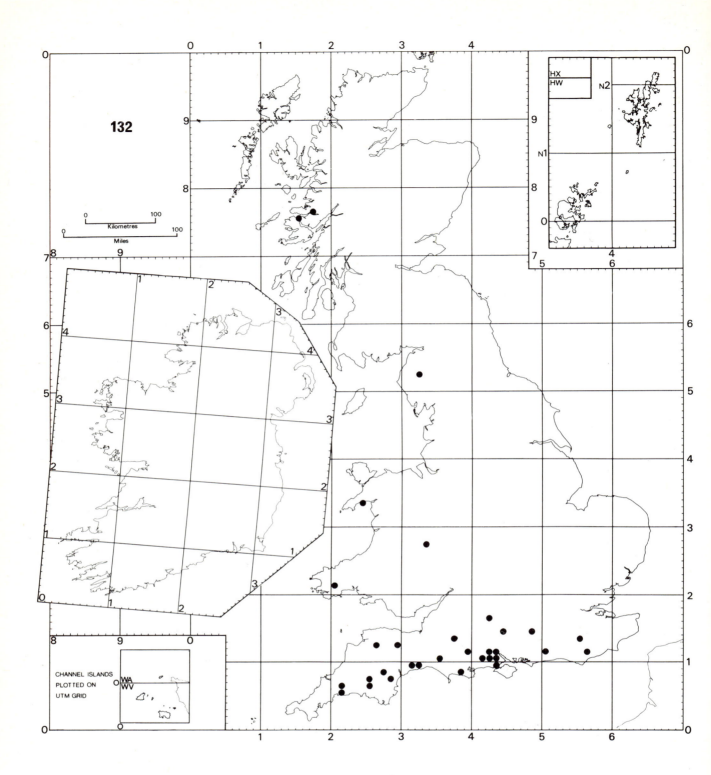

132 *Porina coralloidea* P. James

A recently described species which is an indicator of woodlands of long continuity. It colonises mossy boles of *Quercus*, rarely *Fraxinus* and *Fagus,* in sheltered, shaded and rather humid situations. It is a more or less faithful species of the *Lobarion pulmonariae.* Outside the New Forest, Hampshire, where it is frequent, *P. coralloidea* is restricted to a few ancient parklands in southern England, with a few outliers in Wales, the Lake District and western Scotland; it has not been recorded from Ireland. Abroad it is so far only known from France (Brittany) and the Canary Islands.

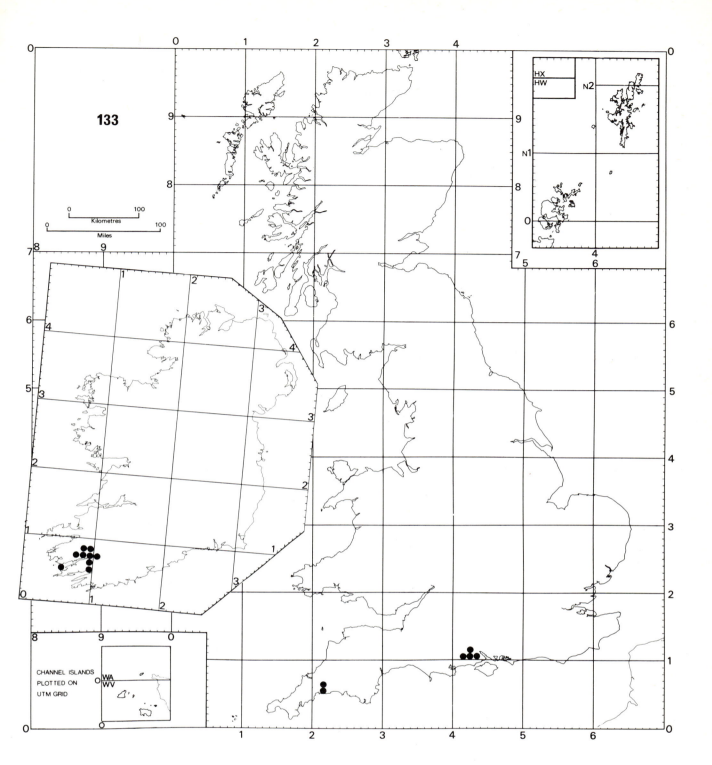

133 *Porina hibernica* P. James & Swinscow

An extremely local species, which colonises mature *Quercus* in sheltered, rather humid ancient woodlands. It tends to grow on smooth bark surfaces between bryophytes, and is faithful to the *Lobarion pulmonariae*. Frequent in many of the ancient woodlands in the New Forest, the only other English records are from Boconnoc Park, Cornwall, where it grows on ancient parkland trees; it is locally frequent in the Killarney Lakes region of south-west Ireland. Abroad it is known from the Azores, and possibly eastern U.S.A.

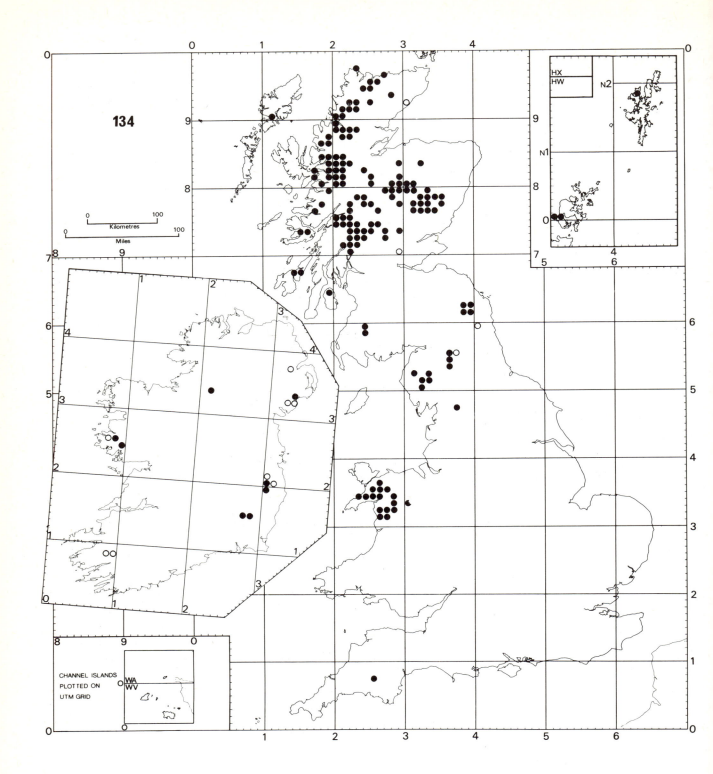

134 *Pseudephebe pubescens* (L.) M. Choisy

A characteristic species of coarse, especially granitic, exposed siliceous rocks in wind-swept upland areas, and a component of the *Rhizocarpetum alpicolae.* Locally abundant in the Scottish Highlands, Snowdonia and the Connemara mountains, but scarce elsewhere. Distinguished from *Bryoria bicolor* (Ehrh.) Brodo & D. Hawksw. by the compact thalli, often forming rosettes, frequent apothecia, and consistantly PD— reaction. There are no immediate threats to its status in the British Isles. An arctic-alpine species, with a very wide circumboreal distribution in both the southern and northern hemispheres.

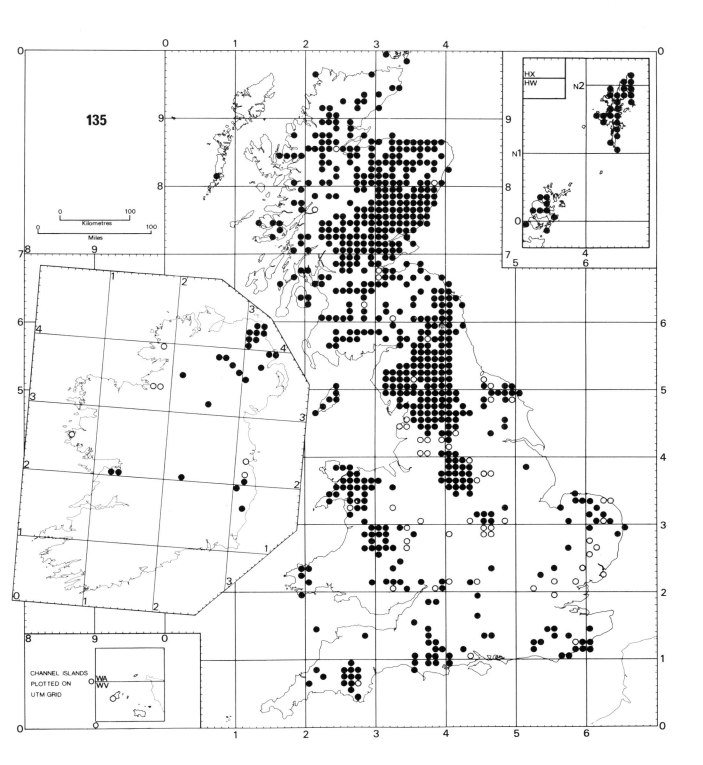

135 *Pseudevernia furfuracea* (L.) Zopf

A characteristic species of exposed, well-lit bark and wood with a pH of 3.0-4.0, occurring mainly on boreal-montane coniferous trees and fence posts over most of its range, but extending onto deciduous trees in moderately polluted areas (absent where mean sulphur dioxide levels exceed about 60 μg m⁻³). Also present on siliceous rocks in similar communities to those in which it occurs on trees, to form the *Pseudevernietum furfuraceae;* this community is the northern and eastern counterpart of the *Parmelietum revolutae* which occurs on bark of pH 5.0-5.5. A morphologically variable species, which also has two chemical races: var. *furfuracea* (med. C−) and var. *ceratea* (Ach.) D. Hawksw. (med. C + red); both are mapped together here but var. *furfuracea* occurs more commonly in the south, decreasing northwards. Formerly regularly fertile in Britain, but rarely so today except in the old Caledonian forests of east Scotland. Largely confined to upland areas of Europe, but with outlying stations (all var. *furfuracea*) in Central America, Bolivia, Ethiopia and the east African mountains.

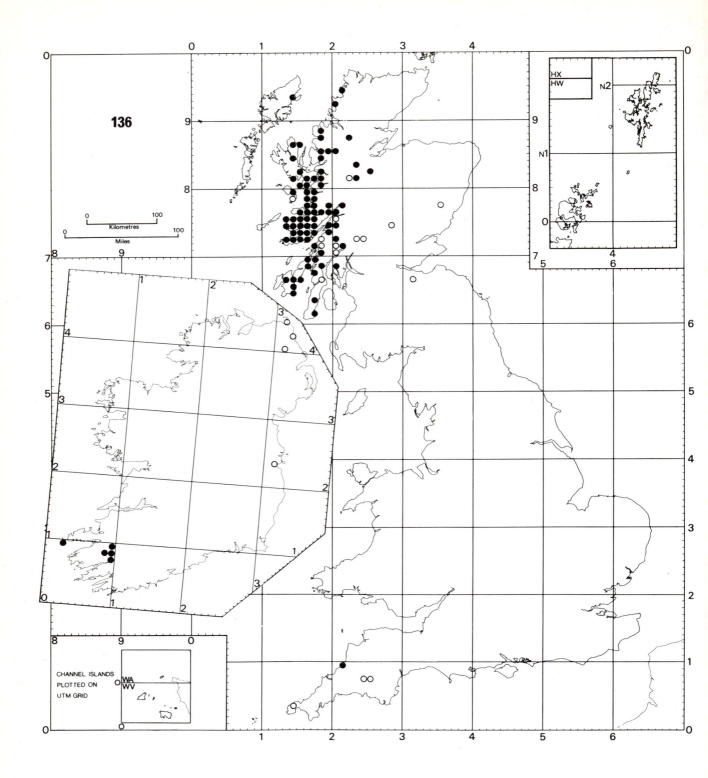

136 *Pseudocyphellaria crocata* (L.) Vainio

A markedly euoceanic species confined to moist, sheltered, well-wooded sites. Like the *P. intricata* complex (map 137), *P. crocata* is a locally frequent species on mossy boles and branches of *Corylus, Fraxinus, Quercus, Sorbus aucuparia,* and in boggy sites, *Salix.* It sometimes occurs on old *Calluna* stems by stream-sides and on mossy boulders in damp woodland and tree-lined margins of freshwater and sea lochs. It appears to require slightly moister conditions than *P. intricata.* It is a member of the *Lobarion pulmonariae,* and is now almost confined to west Scotland and a few sites in west and south-west Ireland; unlike *P. intricata,* it has never been recorded from Wales or south-west Scotland; however, although apparently now extinct on Dartmoor, it still survives at one site on the north coast of Cornwall. Drainage and forestry may have affected this species, reducing its distribution in south-west England and north-east Ireland. Abroad it is world-wide in distribution, occurring in temperate and semi-tropical sites with an oceanic climate, but in Europe it is limited to highly oceanic sites from Norway to Portugal.

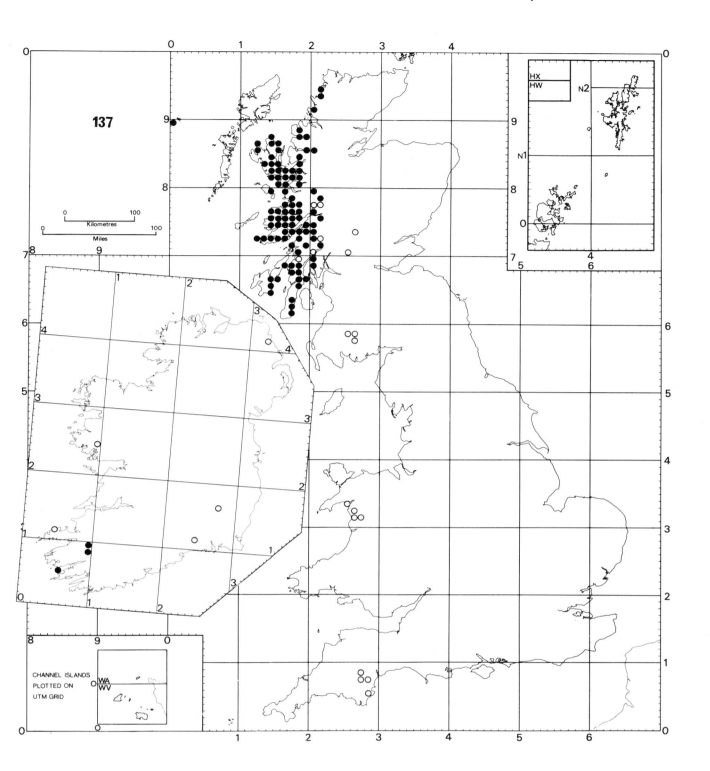

137 *Pseudocyphellaria intricata* (Delise) Vainio, s. lat.

A markedly euoceanic species confined to moist sheltered well-wooded sites, the *P. intricata* aggregate frequently occurs on mossy boles and branches of *Corylus, Fraxinus, Quercus, Sorbus aucuparia,* and in boggy sites, *Salix.* It is rare on old *Calluna* stems near sheltered stream sides, but more frequent on mossy boulders and rock outcrops in damp woodland or sheltered tree-lined margins of freshwater and sea lochs. A member of the *Lobarion pulmonariae,* the *P. intricata* complex is now confined to west Scotland and west and south-west Ireland; in Scotland it is locally frequent on mossy *Corylus* in moderately sunny valleys. It is now extinct in North Wales and south Devon—in these areas forest management and land drainage have contributed to its disappearance. Two species, *P. intricata* s. str. (medulla KC—) and *P. norvegica* (Gyelnik) P. James (medulla KC+ orange - pink) are not distinguished on the map; both have similar distributions. Abroad *P. norvegica* is known only from west Norway; *P. intricata* is worldwide, occurring in temperate and subtropical areas with an oceanic climate, but is limited in Europe to hyperoceanic areas.

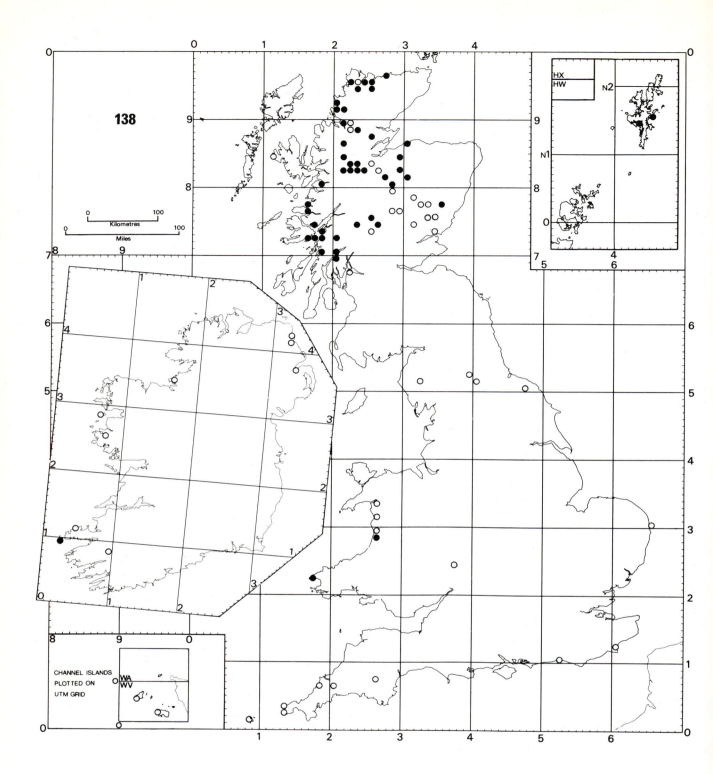

138 *Psoroma hypnorum* (Vahl) Gray

A species of several distinct habitats: (a) on mosses over leached, acid bark at the base of old (sometimes decrepit) trees, especially *Betula* but also *Quercus* in open, derelict woodland; in this habitat it also colonises decaying logs, (b) on sheltered and mossy acid rocks in streams in partially shaded or well-lit situations, (c) on soil amongst mosses on acidic dunes, or leached soils in crevices along sheltered coasts, and (d) amongst mosses, commonly *Racomitrium lanuginosum,* on exposed areas of mountain summits above 900 m. From the diversity of substrates and habitats it is difficult to define the exact ecological requirements of this species in Britain. Its basic need appears to be for leached, acid substrates which tend to dry out rather slowly but are not waterlogged. It now appears to be largely confined to Scotland and south-west Ireland, but it is an easily overlooked species and is probably under-recorded. Abroad it is a circumboreal and montane species, extending into the Arctic Circle and along major mountain chains. It is widespread in temperate and montane areas of the southern hemisphere, where it is also much more variable in form.

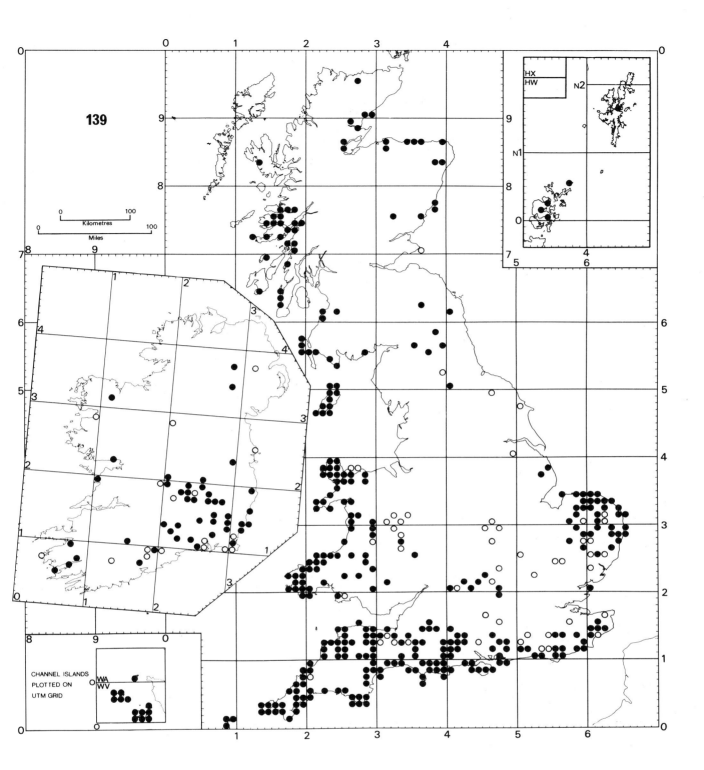

139 *Ramalina baltica* Lettau

A corticolous and saxicolous species preferring sunnier, drier habitats than most other British representatives of the genus, except *R. duriaei*. When corticolous, it occurs mainly on the dry sheltered sides of boles of nutrient-rich, basic-barked (pH 4.5-5.9) trees and shrubs in wind-exposed parklands and hedgerows especially near the coast. It is most frequent on *Ulmus, Fraxinus, Acer pseudoplatanus* and *Quercus,* but may occur on other phorophytes (including old conifers) and palings especially by dusty roads. It occurs on soft or hard rocks, and on these substrates is particularly frequent on the coast where it grows in sheltered, mainly dry recesses often facing the sea. In inland sites, it is commonly present on the sheltered side of church towers and gravestones. When on trees it is included in a dry facies of the *Ramalinetum fastigiatae;* when saxicolous it has affinities with a dry facies of the *Ramalinetum scopularis. R. baltica* is extremely sensitive to pollution from sulphur dioxide and inorganic fertilizers, and its absence now from central and most of northern England is probably due to this cause, also tending to accentuate its lowland, coastal distribution in the rest of England, Wales and Scotland; it is probably a widespread species in central and eastern Ireland. In Europe it is continental, extending north to Norway and eastwards through the Baltic to western Russia. *R. baltica,* under current concepts, consists of two races: that in western Europe contains divaricatic acid, and that in eastern Europe evernic acid. In western Europe the species appears to be very close to *R. canariensis* and may in fact be conspecific; if this is so, the identity of the east European plant may need further study. The two races converge in Sweden, and the eastern race has not been reported from the British Isles.

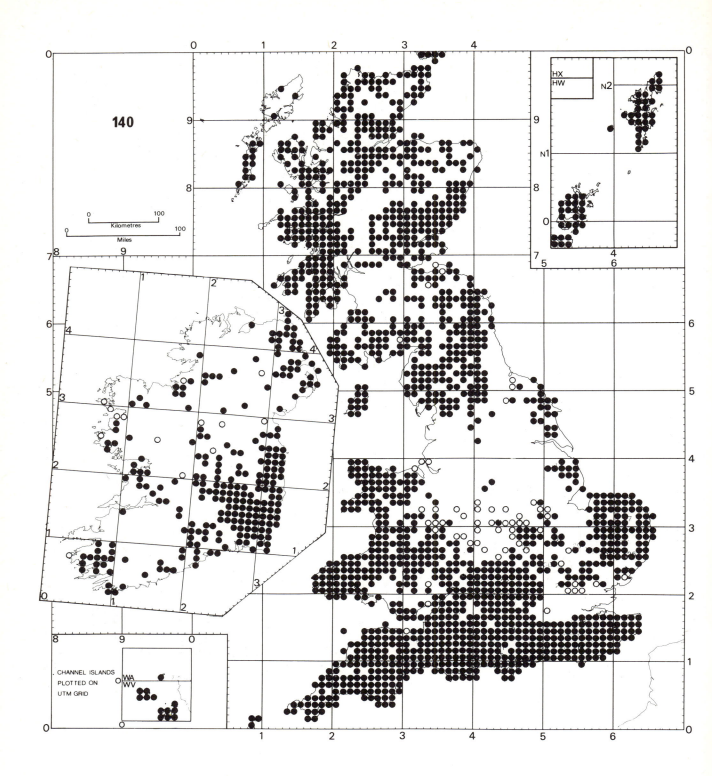

140 *Ramalina farinacea* (L.) Ach.

A common and widespread corticolous species of partially shaded to well-lit, mainly wayside, hedgerow or parkland shrubs and trees with moderately nutrient-rich acid or basic barks (pH 3.8-5.8). *R. farinacea* grows on a wide range of phorophytes, as well as palings, reflecting its wider ecological amplitude than other British species. It is also less sensitive to sulphur dioxide pollution (up to 60 μg m^{-3}) and inorganic fertilizers, but their influence still accounts for its disappearance from north-central England, the London area and the central Lowlands of Scotland. The species is also thinly distributed in the Scottish Highlands. Elsewhere *R. farinacea* is a frequent species, especially in windy wayside situations, but it also penetrates denser woodland, occurring especially in the upper tree canopy; it is widespread in Ireland. Though primarily assigned to the *Ramalinetum fastigiatae* this species, because of its wide ecological amplitude, occurs as an accessory, but sometimes important, species in other related associations such as the *Parmelietum revolutae* and the *Teloschistetum flavicantis*. A number of chemical races occur in this species—no differences in distribution and ecology of these have been discerned. It is widespread throughout Europe.

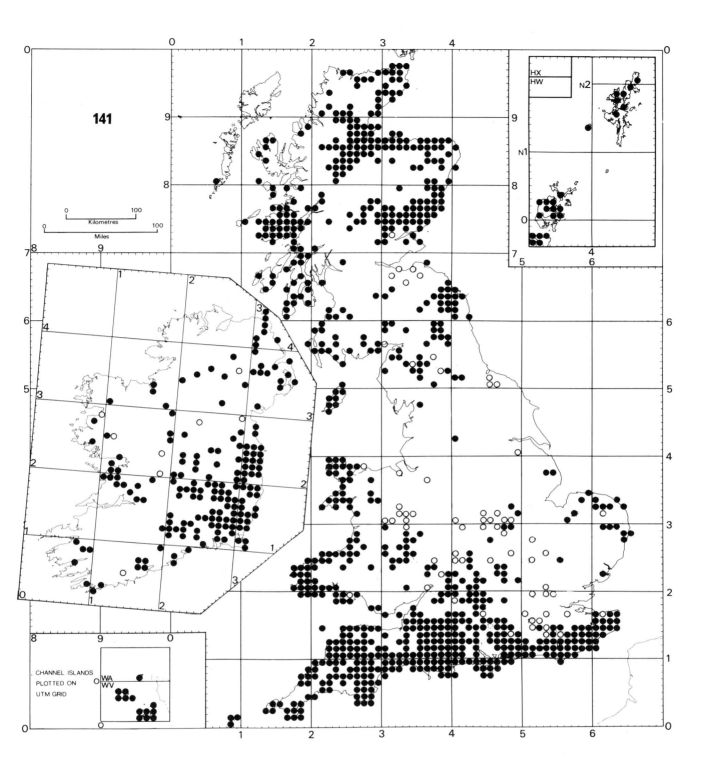

41 *Ramalina fastigiata* (Pers.) Ach.

A corticolous species on well-lit, wind-exposed, mostly wayside hedgerow or parkland trees, with a nutrient-rich, basic bark (pH 4.3-5.9), particularly *Ulmus, Acer pseudoplatanus, Fraxinus,* orchard trees and, near the coast, *Prunus spinosa;* it may occur on other phorophytes and palings, particularly when these are enriched by airborne organic nutrients. Very rarely, it grows on softer, sub-basic rocks of church towers and gravestones. Though common on boles, it is also a twig species. It is an important element of the *Ramalinetum fastigiatae.* Very sensitive to air pollution, it cannot tolerate inorganic fertilizers or a mean winter SO_2 level of more than 35 μg m^{-3}, and has therefore largely disappeared from central and northern England. Elsewhere it shuns upland areas with high rainfall, and is most common on or near windy coasts or hedgerows in exposed, low, hilly terrain. It is found throughout central and eastern Ireland. Abroad it is a widespread, temperate hemiboreal species, preferring coastal regions.

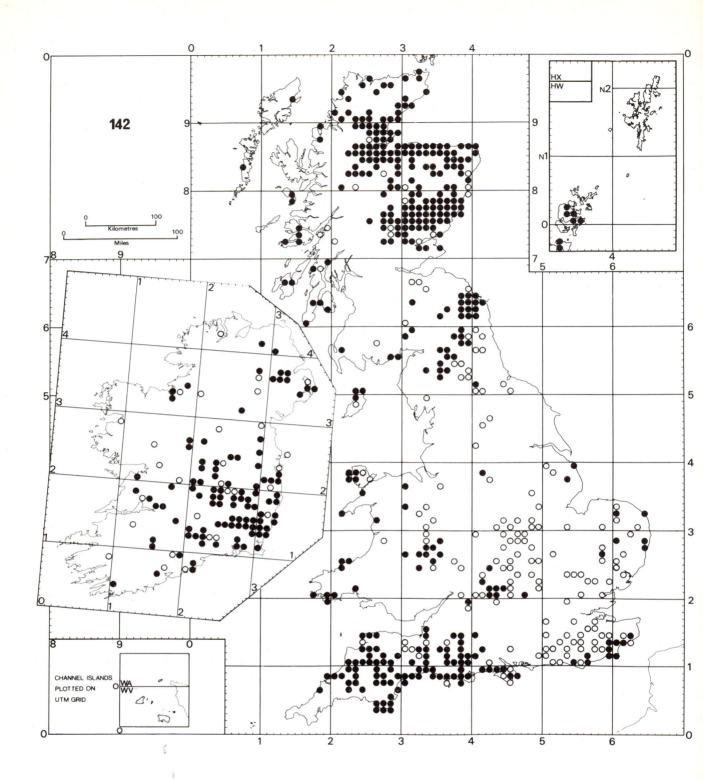

142 *Ramalina fraxinea* (L.) Ach.

An elegant corticolous species on well-lit, wind exposed, mostly wayside, hedgerow or parkland trees with a nutrient-rich, basic bark (pH 4.7-5.8), particularly *Ulmus, Acer pseudoplatanus* and *Fraxinus,* and more rarely *Quercus* and *Populus. R. fraxinea* is included in the *Ramalinetum fastigiatae,* and is most well-developed on the south and west facing aspects of tree boles or on inclined main branches. The species is very sensitive to pollution, decreasing in areas with mean winter SO_2 levels exceeding 35 μg m^{-3}, and is, for this reason, extinct or very rare in most of central and south-east England where it was formerly locally abundant; its range has probably been further reduced by the effect of Dutch Elm Disease on its principal phorophyte in these areas. Elsewhere it is most abundant in south-west England and north-east Scotland and over most of central and eastern Ireland; it is mainly absent from high-rainfall, forested areas in the west of Britain. Farm sprays, inorganic fertilizers and felling of wayside trees will almost certainly reduce its distribution in Britain still further. It is a widespread species in continental Europe, but rarer in the Mediterranean zones.

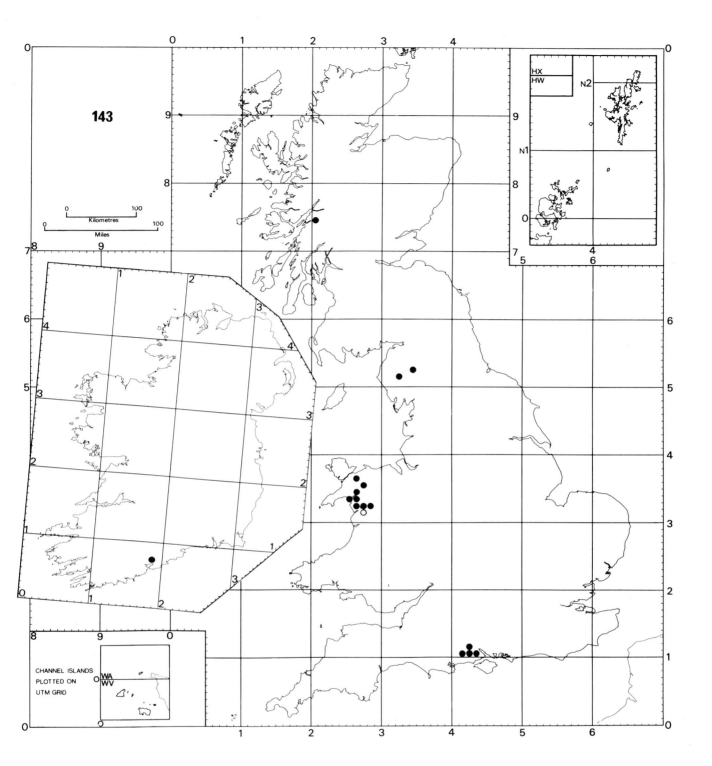

143 *Rinodina isidioides* (Borrer) H. Olivier

A very local corticolous species occurring on the boles of mature *Quercus* in a few sheltered old woodlands of long continuity. It generally occurs amongst mosses on partially sun-lit boles of trees on the south and west margins of woods, although it can tolerate more shaded conditions within such woods. It is a member of the *Lobarion pulmonariae*, restricted in Britain to two principal areas, the New Forest and North Wales (especially around Harlech, Arthog and Dolgellau); outliers occur in the Lake District and western Scotland. There is a single locality for south-west Ireland. Distinguished from other species in the genus by the grey-white, isidia-like lobes; occasionally fertile. Not widely known abroad, but occurs in France (Brittany).

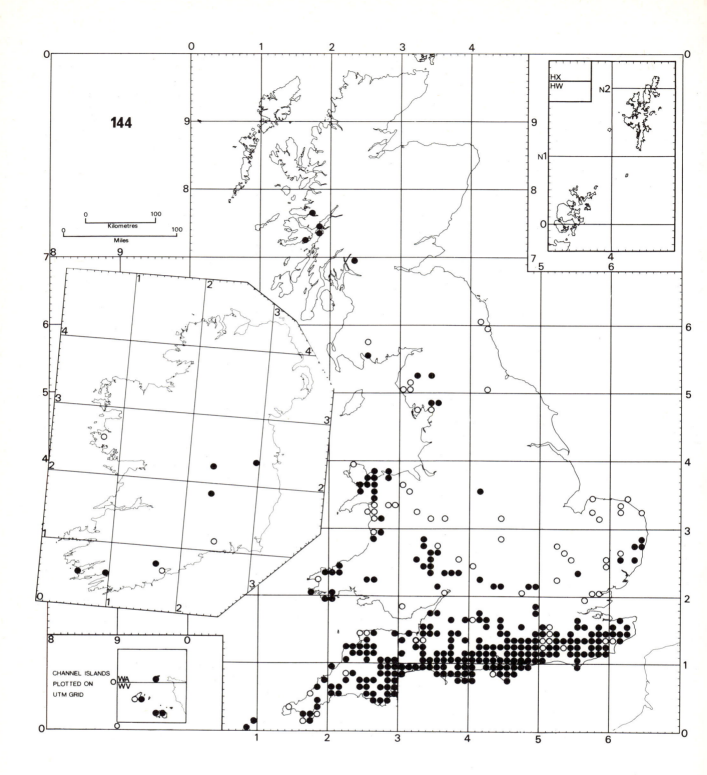

144 *Rinodina roboris* (Dufour ex Nyl.) Arnold

A characteristic species of sunny, well-lit, rough, more or less nutrient-rich bark, of medium aged to old trees, especially *Quercus, Ulmus* and *Fraxinus,* in margins (usually south and west facing) of woods, waysides and hedgerows. It is often frequent on trees in fields not affected by inorganic fertilizers. Its scarcity in central and eastern England is probably due to sulphur dioxide pollution and modern farming methods. In southern England it is far commoner than elsewhere in Europe, and occurs in the facies of the *Lobarion pulmonariae* requiring well-lit conditions, the *Physcietum ascendentis* and *Parmelietum revolutae*. In northern England it occurs very rarely on isolated trees in sheltered sunny sites, or in parkland, usually on basic or rich alluvial soils. In Ireland it is mainly in the south-west (around Killarney), but extends eastwards where suitable habitats exist. It is a predominantly southern species, which reaches the northern limit of its range in north-west Scotland. In Europe it is confined to the oceanic areas in the west, from Brittany south into Portugal.

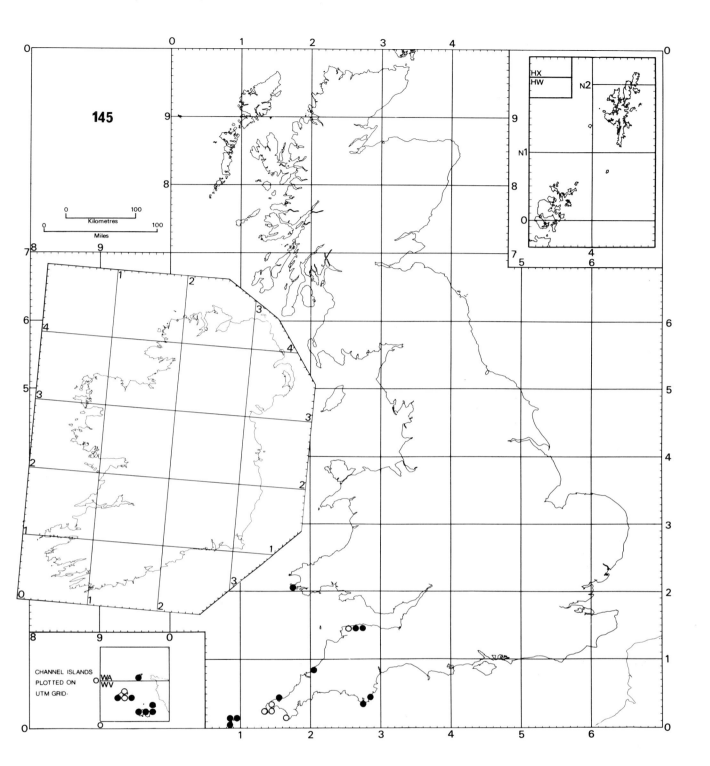

145 *Roccella fuciformis* (L.) DC.

An elegant species of dry, vertical or recessed east- or north-facing rocks not directly wetted by rain, in sunny, mostly frost-free coastal sites. Rarely on the dry sheltered side of old trees near the coast. Xeric-supralittoral, preferring hard (commonly granite) rocks, it belongs to the *Sclerophytetum circumscriptae,* a community which includes several taxa like the present species, at the northern limit of their range. Locally frequent in the Channel Islands and Isles of Scilly, the most northerly record for this species in Europe is the island of Skomer, Pembrokeshire; it is surprisingly absent from Ireland. Outside Britain it is known from western France, Spain, Portugal, the Canary Islands and the Azores.

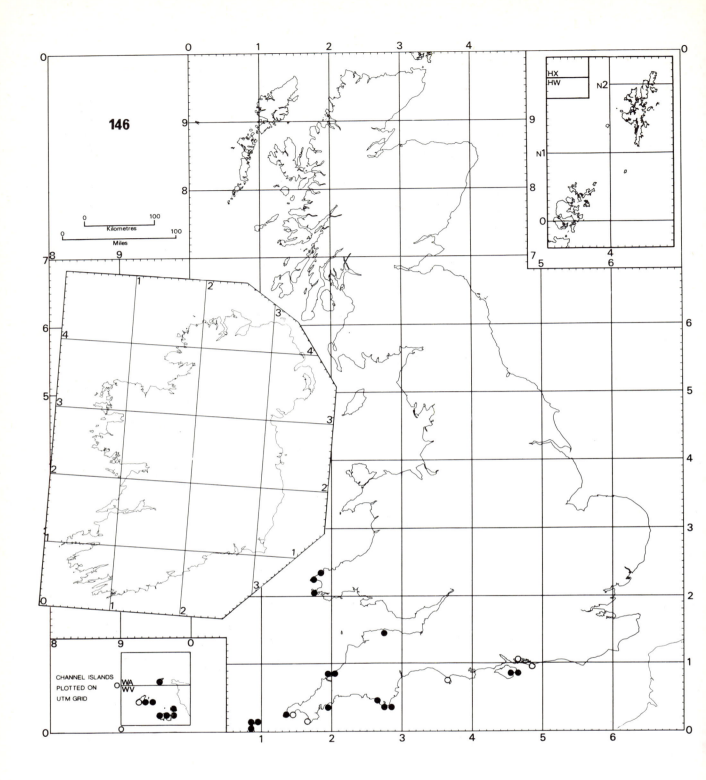

146 *Roccella phycopsis* (Ach.) Ach.

A predominantly saxicolous species of dry vertical or recessed rock faces, particularly those of east- or north-facing aspect; also occasionally on the east and north sides of church towers and on the driest side of old trees. *R. phycopsis* is a predominantly coastal species, in the xeric-supralittoral zone, which belongs to the *Sclerophyton circumscriptae* and often occurs with *R. fuciformis.* It is perhaps more frequent than *R. fuciformis,* formerly extending further eastwards to Sussex and still present on church towers on the Isle of Wight. Common in suitable coastal sites on the Channel Islands and Isles of Scilly; it is surprisingly absent from Ireland. Abroad it is known from western France, Spain, Portugal, north Mediterranean, the Canary Islands and the Azores.

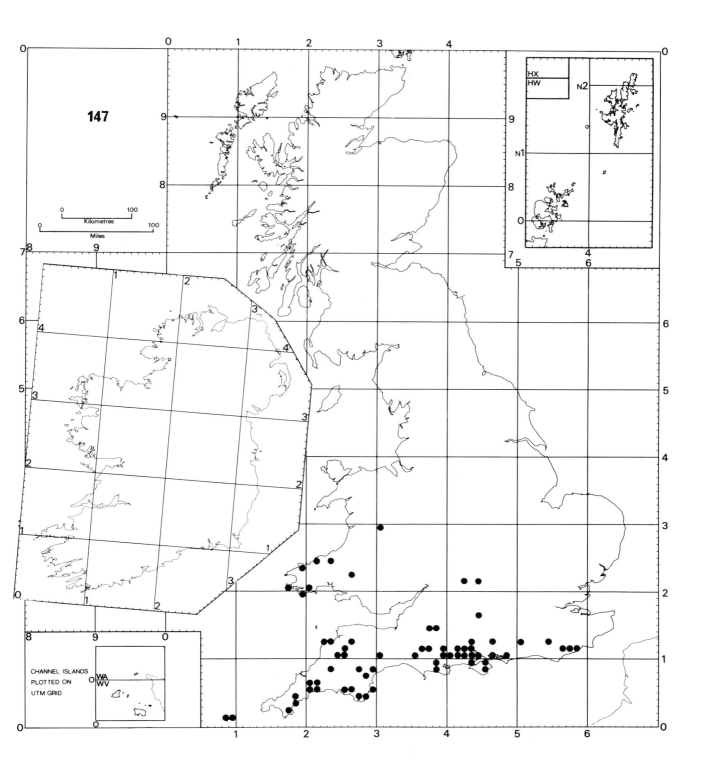

147 *Schismatomma niveum* D. Hawksw. & P. James

A recently described old woodland indicator species, mainly growing in sheltered, poorly-lit dry bark recesses at the bases of old mature *Quercus*. In old woodland areas *Schismatomma niveum* is a frequent member of the *Lecanactidetum abietinae*. This species can be readily separated from *Haematomma ochroleucum* var. *porphyrium*, *Ochrolechia turneri*, etc., by the K—, C— and PD+ deep yellow reactions of the thallus. It is local and widespread in southern Britain, but has not yet been recorded from Ireland. Otherwise known from France (Brittany and Normandy).

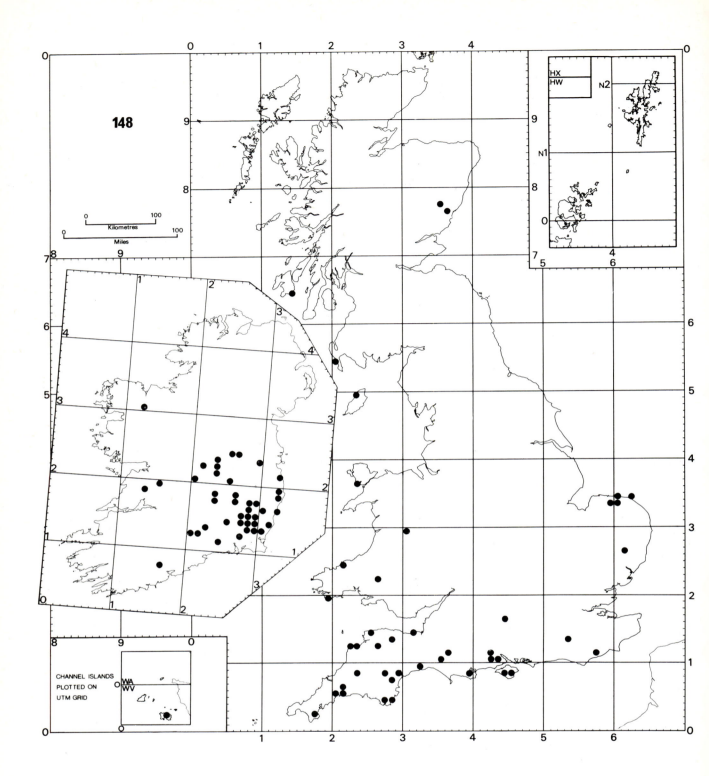

148 *Schismatomma virgineum* D. Hawksw. & P. James

Recently described, this corticolous species, characteristic of dry, moderately shaded bark of ancient trees, mainly *Quercus* but also *Fraxinus* and *Salix,* forms extensive patches. It prefers open parkland, or more rarely close-canopy woodland, in areas with low sulphur dioxide pollution (35-40 μg m^{-3}) and is absent from hypertrophicated tree barks. It is found as a component of some facies of the *Lecanactidetum premnae.* In the field it is brilliant white, and, unlike *S. niveum* (map 147), has consistently K+ pale yellow and PD— reactions. It is much more widespread than *S. niveum,* extending into East Anglia, North Wales, Scotland and Ireland. In Europe it is known from France (Brittany and Fontainebleau) and Denmark.

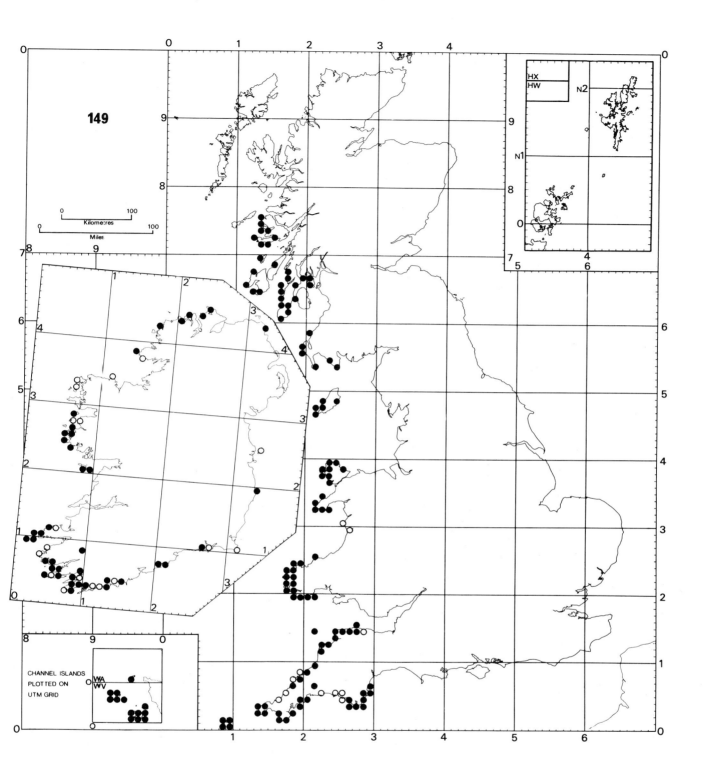

49 *Sclerophyton circumscriptum* (Taylor) Zahlbr.

Locally common in crevices and beneath overhangs on siliceous rocks, but restricted to islands and peninsulas in western Britain. It is usually maritime, in the xeric-supralittoral zone, and always in shade. Rarely it can be encountered in the terrestrial zone on sheltered rock faces. It appears to prefer warmer situations, but is intolerant of wetting by rain or spray. Nevertheless, sea-water is apparently vital to it in Britain. The apparent absence of this species from a number of sites in south-west England may reflect recent under-recording. It is difficult to recognise in the field since it resembles a sterile white crust; however, it is always, albeit inconspicuously, fertile. Found near sea-coasts in western Europe, from Brittany to Portugal.

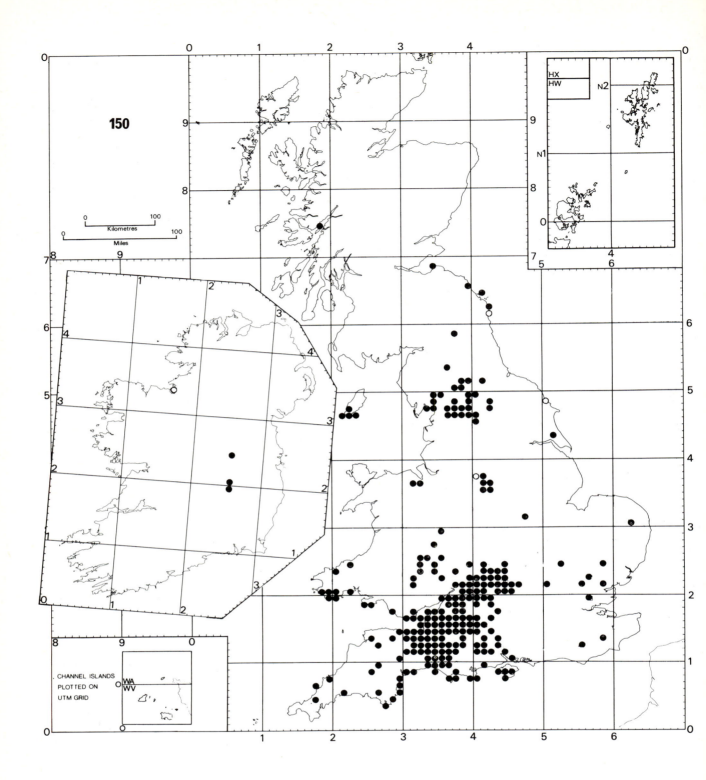

150 *Solenopsora candicans* (Dickson) Steiner

A saxicolous species of basic substrates, colonising both man-made materials, such as gravestones and wall tops, as well as pure, moderately soft to hard limestones; it does not occur on chalk. It is particularly frequent in Somerset, Worcestershire and Herefordshire, where it grows in most churchyards as well as on buildings and exposures of oolitic limestone in the Cotswolds. Elsewhere it is not uncommon on the Carboniferous limestones of northern England (north Yorkshire and the Peak District); outside these areas an affinity for coastal churchyards, or limestone outcrops on or near the coast, seems apparent from the map. Its most northerly station is on the island of Lismore, Argyllshire. It favours the moister facies of the *Caloplacetum heppianae* and the *Placynthietum nigri,* and tends to grow on slightly sheltered, moisture-retaining substrates or on the flat upper surfaces of chest-tombs in sheltered, but not heavily shaded churchyards. In Europe it is widely distributed, but is essentially a western and Mediterranean species; also recorded at least from North America.

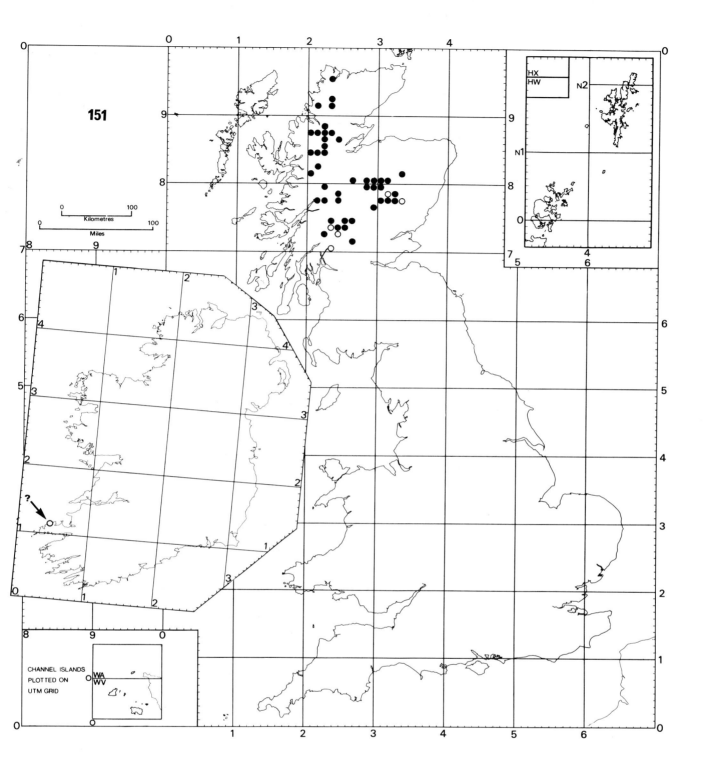

51 *Solorina crocea* (L.) Ach.

A calcifuge terricolous species characteristic of leached moraines, solifluction areas and rock crevices, where there is late snow-lie and little other vegetation. A component of the *Gymnomitreto-Salicetum herbaceae,* often forming extensive swards in such areas. More rarely associated with mica-schists. Probably confined to the Scottish Highlands, mainly above 900 m, but there is one old record from Co. Kerry, Ireland descending to about 600 m; found especially on wind-swept ridges. Easily recognised by the bright orange underside; it is almost always fertile. Status only likely to be affected by habitat destruction (e.g. by increased recreational pressure). A circumpolar, northern boreal to high-arctic species in the northern hemisphere, with a few sites in the middle boreal zone, extending south to the Alps and Pyrenees in Europe. Also reported from Japan and New Zealand, but not from Australia, South America or Antarctica.

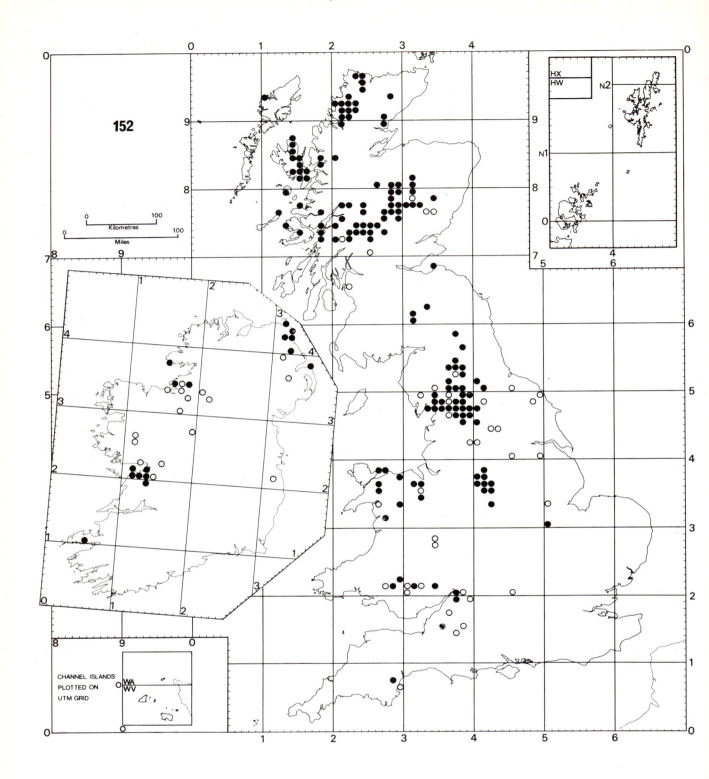

152 *Solorina saccata* (L.) Ach.

A strictly calcicole species most commonly found in moist, sheltered crevices and on ledges of hard limestone outcrops and also on other calcareous rocks; usually amongst mosses, and associated with species of the *Caloplacetum heppianae*, *Gyalectetum jenensis,* or *Leproplacetum chrysodetae* on adjacent rock. Sometimes found in short calcareous turf and on old walls and calcareous sand-dunes. In the British Isles its distribution is mainly limited to areas with hard limestones, ranging from sea level to 1200 m (on Ben Lawers). Although the species has some invasive tendencies, experience in the Peak District suggests that it behaves as a relict species, being almost entirely confined to sites with a long history of ecological continuity. Status only likely to be endangered by habitat destruction. An arctic to temperate circumpolar species, widespread in Europe and North America, becoming montane in the southern parts of its range, and apparently absent from the southern hemisphere, except South Africa.

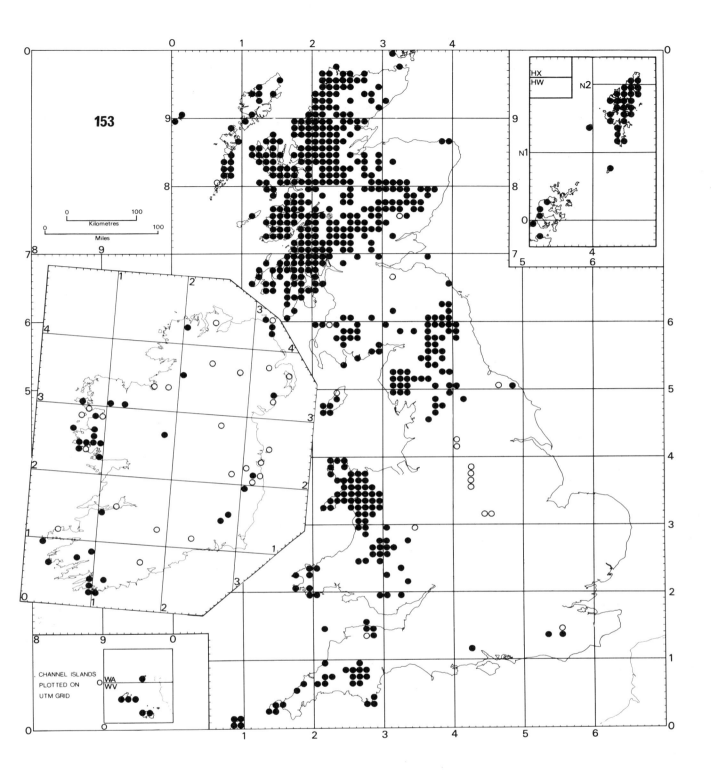

53 *Sphaerophorus globosus* (Huds.) Vainio

A widespread and often common species in most rocky moorland and upland areas of Britain. It is primarily saxicolous, often amongst other lichens and mosses, but in areas with more than 150 cm rainfall per annum it may extend up the boles of trees, especially *Betula* and *Quercus* in leached, moist woodlands; it is often terricolous on moorland. It enters several important associations: the *Parmelion laevigatae* of leached wet upland woods, a facies of the *Cladonietum coniocraeae* in a few, drier lowland woods, the saxicolous facies of the *Pseudevernietum furfuraceae* and, above all, the *Parmelietum omphalodis.* In Britain it is widely distributed, often abundant in the Scottish Highlands and ascending to an altitude of 1200 m; it is probably common and widespread in Ireland, less common but widely distributed in northern England, and frequent in North Wales. Elsewhere it is rather local, restricted by the availability of suitable habitats; it has completely disappeared from the Midlands and almost so from south-east England due to air pollution, drainage and habitat modification. A temperate circumboreal species, with outliers in the southern hemisphere.

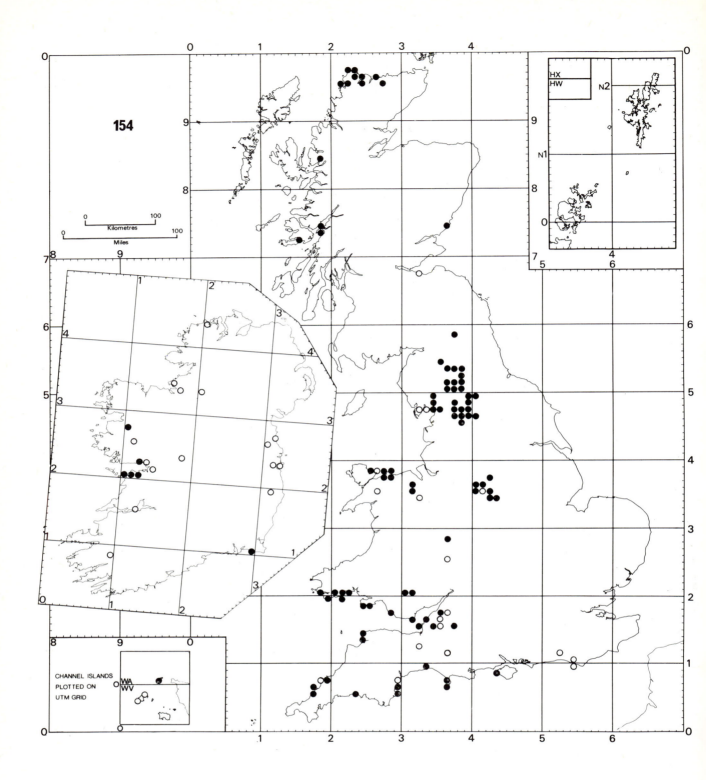

154 *Squamarina cartilaginea* (With.) P. James

A markedly calcicole species colonising a range of hard calcareous substrates. Whilst it will tolerate relatively exposed sites, it prefers some shade, often entering a wet, sheltered facies of the *Caloplacetum heppianae* and a mossy woodland facies of the *Placynthietum nigri*. It mainly colonises crevices on limestone cliffs or the summits of limestone pavement, often spreading to adjacent rocks; it is rare on basic igneous rocks, such as calcareous basalt. There are a number of records for this species on calcareous dunes (pH *c.* 8.0) derived from limestone outcrops or shell-sand. Its distribution follows that of hard limestones, especially Carboniferous limestone; it is rarer on Jurassic limestones and there are very few records from chalk. It occurs from sea level to *c.* 500 m. There are two chemical races: var *cartilaginea,* which is confined to coastal areas, contains psoromic and conpsoromic acid, and var *pseudocrassa* (Mattick) D. Hawksw., on both coastal and inland sites, is without these acids. It occurs throughout the Mediterranean, extending into north Africa, Asia Minor and the Ukraine; also present in the Canary Islands, the Azores, and possibly North America.

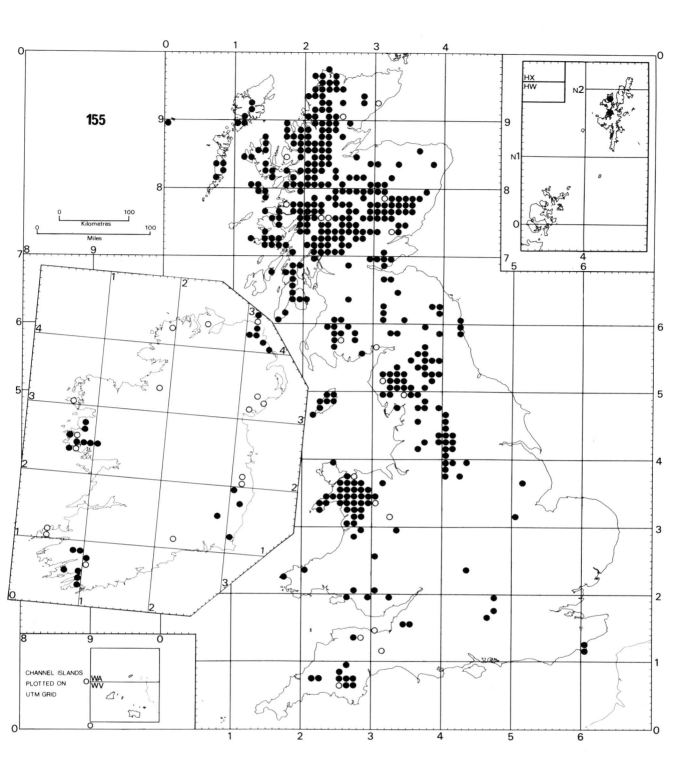

55 *Stereocaulon vesuvianum* Pers.

This species is characteristic of moderately to well-lit hard siliceous (particularly igneous) rocks with a high metal content, occurring in the facies of the *Acarosporetum sinopicae;* also colonising gritstone walls and even brick in areas where these substrates have become enriched by heavy metals, and exceptionally occurring on worked wood. The species appears to be tolerant of sulphur dioxide and is in process of extending its range in lowland England on man-made substrates; it was known only from areas with igneous or metamorphic rocks in the last century. Increasing lead pollution can be expected to favour the expansion of both this species and *S. pileatum. S. vesuvianum* is a widely distributed and polymorphic species, occurring in both hemispheres, especially in volcanic regions, extending from the antarctic to the arctic zone.

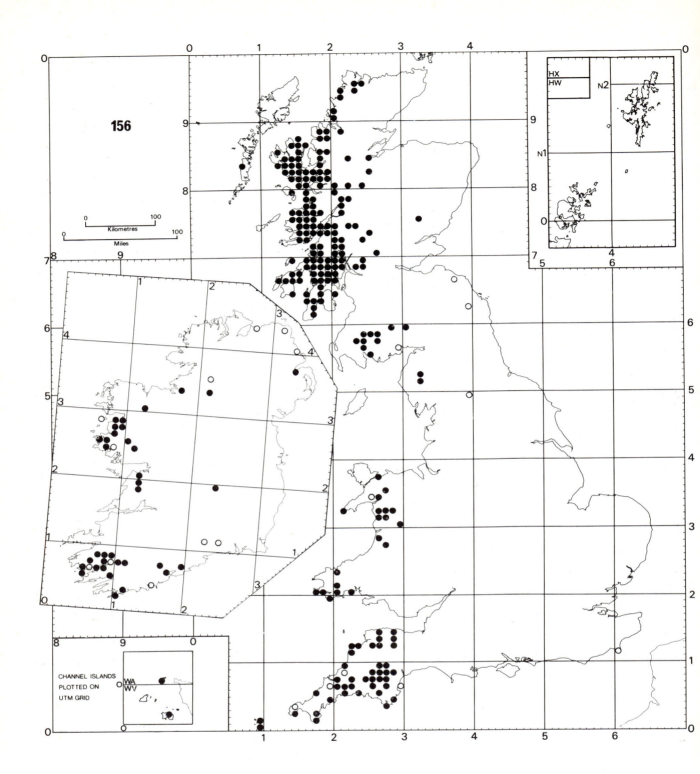

156 *Sticta fuliginosa* (Dickson) Ach.

An old woodland indicator species, confined to ancient oceanic woodlands where it occurs on mossy trees and damp rocks in humid, sheltered situations. A member of the *Lobarion,* but more oceanic than *S. sylvatica* and *S. limbata* which extend much further east in Britain. Confined to areas where rainfall exceeds 95 cm per annum and to more basic substrates; it is also highly intolerant of acid rain and sulphur dioxide pollution. Very common in the western Highlands of Scotland, locally frequent elsewhere on the western side of Britain south to Cornwall and Devon, but absent now in central, east and south-east England; there is a single nineteenth century record for Dungeness in Kent on *Prunus spinosa* on the shingle beach. Shows little evidence of a broad decline over the last 100 years, but has certainly become more restricted locally with the destruction of old woodlands. Extremely rare in the fertile state. Even more oceanic in Europe than *S. sylvatica;* from west Norway to Brittany and the western Pyrenees. Elsewhere known from Africa, S. America, Australia and New Zealand.

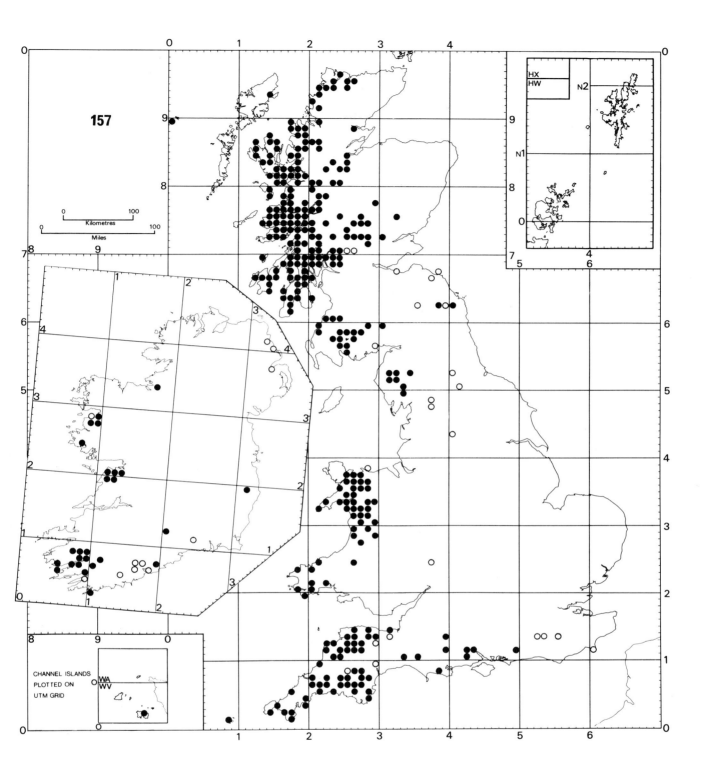

57 *Sticta limbata* (Sm.) Ach.

A sorediate, never fertile, species of bark (and more rarely mossy rocks) in old woodlands, forests and parklands; an excellent old woodland indicator species except in north-west Scotland, where in the moist, clean air of this region, it can be found on young trees in woodlands of more recent origin, and even on roadside trees. It is a characteristic member of the *Nephrometum lusitanicae* association of the *Lobarion* alliance in the British Isles, occurring on mossy bark usually in the pH range 4.5-6.0, and only in areas with mean winter sulphur dioxide concentrations below about 30 μg m⁻³. Formerly more widespread but now confined to the south and west of Britain from the New Forest to Cornwall and northwards to Sutherland, (except for a relict locality on a single old *Fraxinus* in a valley in an ancient wood in west Sussex). It is still quite frequent in Devon, Cornwall and North Wales, common in western Scotland, and widespread in Ireland in the few areas where old woodland survives. Although generally decreasing in England, some New Forest populations have markedly increased in size in the last eight years, perhaps indicative of falling air pollution levels there. On the Continent, it has shown a similar reduction in range over the last 50 years: though locally still plentiful in western Norway, it has now disappeared from Denmark, much of southern Sweden and the rest of the northern European plain, appearing again in the Ardennes and in France from western Normandy and Fontainebleau southwards. In southern Europe it is confined to montane forests (on broad-leaved trees and *Abies alba*), but even here its sub-oceanic distribution is revealed in that its eastern limit now seems to be in *Castanea* forests of Provence, unlike some other members of the *Lobarion* which extend through Italy and Yugoslavia to Greece and beyond. Also known from the Azores, Canary Islands, north Africa, the west coast of North America, and Australasia.

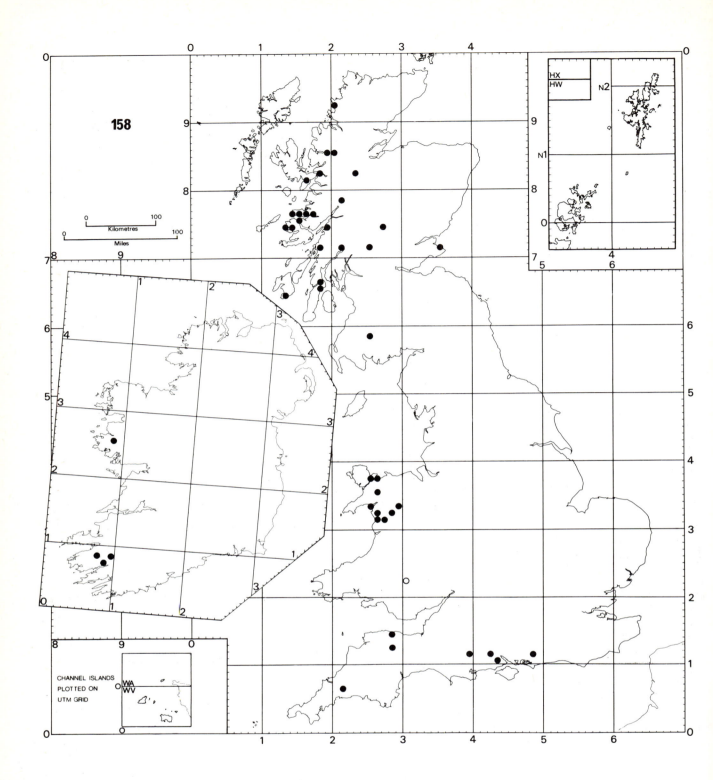

158 *Strangospora ochrophora* (Nyl.) R. Anderson

An inconspicuous, but very distinctive, species colonising more or less nutrient-rich, basic bark of medium aged to old trees, especially *Ulmus, Fraxinus,* and occasionally *Quercus* and *Corylus,* on rich soils. The species prefers the summits of bark corrugations amongst stems of *Frullania,* on the sheltered, not too wet, sides of phorophytes. Although usually preferring sheltered woodland, it may occur on isolated wayside trees near the coast, where it benefits from on-shore sea-mists. *S. ochrophora* is a species faithful to the *Lobarion pulmonariae.* It is widespread but local in Britain; in southern England it is a relict species confined to a few old woodlands and parklands. The species is more frequent in north Wales and lowland areas of western Scotland, where it is widespread, occurring on wayside as well as woodland trees. Reported from parts of southern, central and northern Europe, and also North America, but the present status of the species in many areas is uncertain.

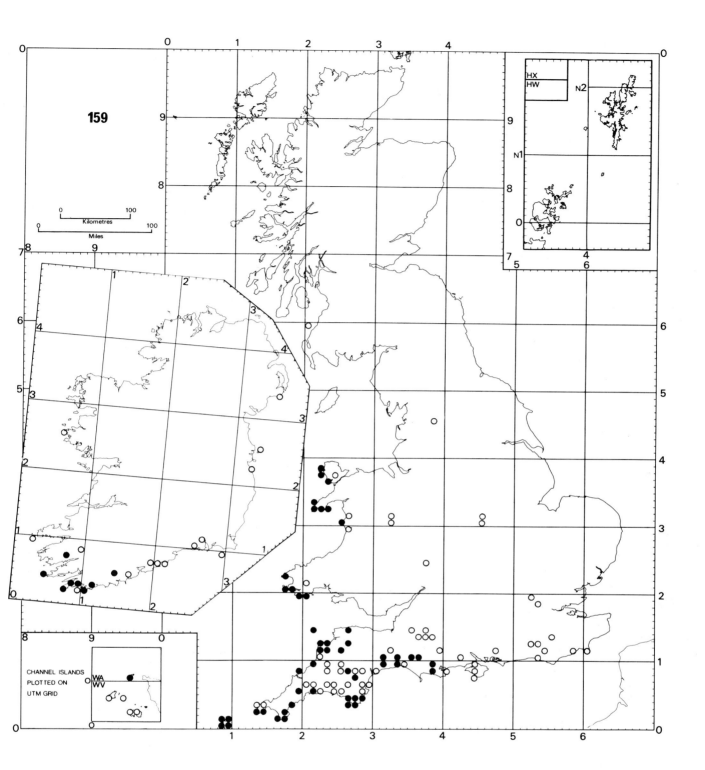

9 *Teloschistes flavicans* (Swartz) Norman

On siliceous rocks, soil, lignum, twigs and bark. A rare species, now restricted to south-west England and Wales. At many sites it is coastal, extending from the xeric supralittoral into the terrestrial halophilic region, on rocks or on soil, where it entwines angiosperm stems, especially *Calluna* and *Festuca* species. In the south-west it is also found at inland stations, on similar substrates, or on nutrient-rich tree trunks (especially of *Acer pseudoplatanus* and *Fraxinus excelsior*) and twigs; on trees it is part of a distinctive community rich in scarce *Physcia* species, the *Teloschistetum flavicantis*. It requires well-lit and moderately wind-exposed situations. *T. flavicans* was formerly much more widely distributed, extending into central England on trees. Its disappearance is attributable to loss of habitat, especially nutrient-rich bark, burning, over collection and air pollution. It is sterile in Britain. This is a well recorded, unmistakeable species. Cosmopolitan in the tropics and warm temperate regions, including Atlantic and Mediterranean Europe from Brittany southwards; extending to alpine or semi-arid regions in some continents. It reaches a northern limit to its distribution in Britain. The world ecology is very variable; it is found on trees and rocks in a wide variety of habitats; also a common component of subtropical lichen communities, where it can be found in dry and dusty situations.

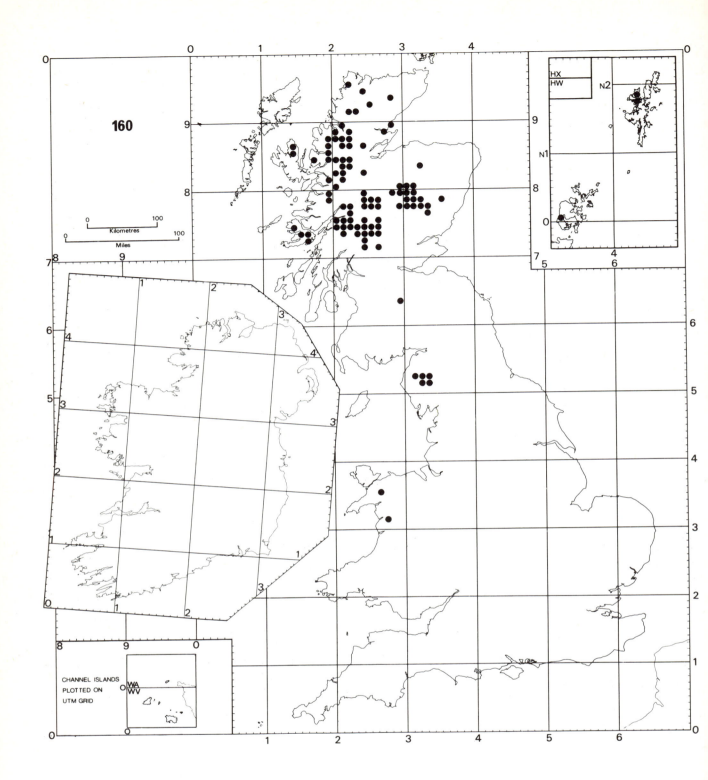

160 *Thamnolia vermicularis* var. *subuliformis* (Ehrh.) Schaerer

An arctic-alpine species of short, exposed, wind-clipped upland heaths, most commonly growing amongst *Racomitrium lanuginosum,* often intermixed with ericaceous shrubs or *Salix herbacea.* Mainly on montane plateaux and summits over 850 m high, but exceptionally descending to almost sea level on dunes in Scotland. The pure white, wiry, usually unbranched and decumbent hollow thalli are unlikely to be confused with any other species. Fruits are unknown, but several lichenicolous fungi are able to grow on this species. *T. vermicularis* var. *vermicularis* has not been correctly reported from the British Isles. This species has an exceptionally wide distribution in montane situations in both hemispheres, although it is apparently absent from Africa. It has been suggested, on phytogeographic evidence, that the species arose in Permo-Triassic times.

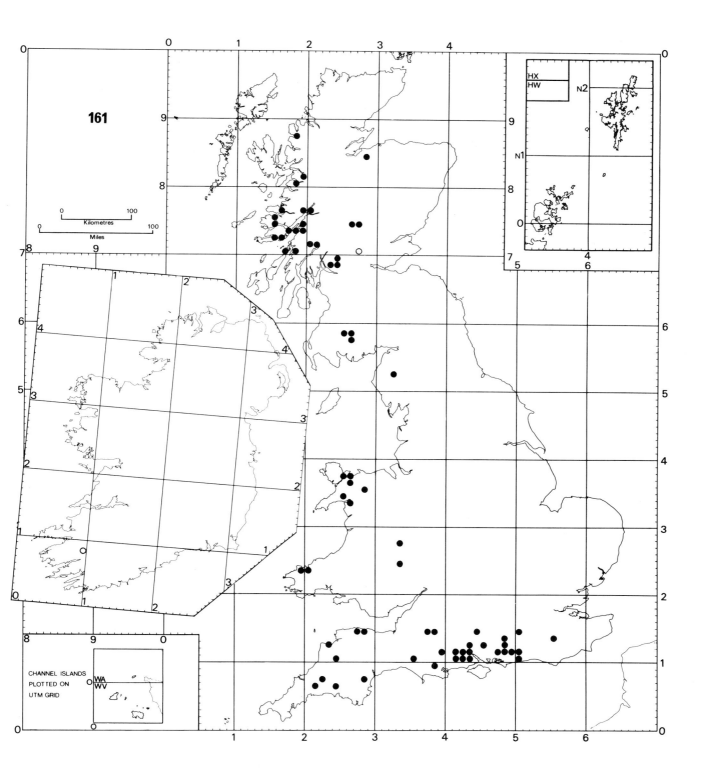

161 *Thelopsis rubella* Nyl.

A rather inconspicuous corticolous species, only recognisable when fertile. Mainly on old *Quercus*, *Fagus* and *Fraxinus* and other trees with less acid bark (>pH 5.0). It is an old woodland indicator species of more open patches within the *Lobarion* community, particularly in grooves and furrows on more water-retentive bark. It shows an "L-shaped" distribution in Britain, characteristic of many sub-oceanic old woodland species. Commonest in the New Forest, and occasional to locally frequent in the west and central Highlands of Scotland, down the west side of Britain and then eastwards to east Sussex along the south coastal zone. It was largely overlooked until recently in the British Isles due to its inconspicuous appearance in the sterile state, being known only from single sites in Scotland and south-west Ireland until 1968; only since then has its distribution been worked out, so it is impossible to say if it has declined, but it is almost certain that it was far more widespread in the past. Widely distributed, but rare, throughout much of Europe, in old forests from west Norway and Denmark through France and north Italy to Czechoslovakia, etc. In the north and west it is a lowland species, but in south Europe its habitat is montane woods.

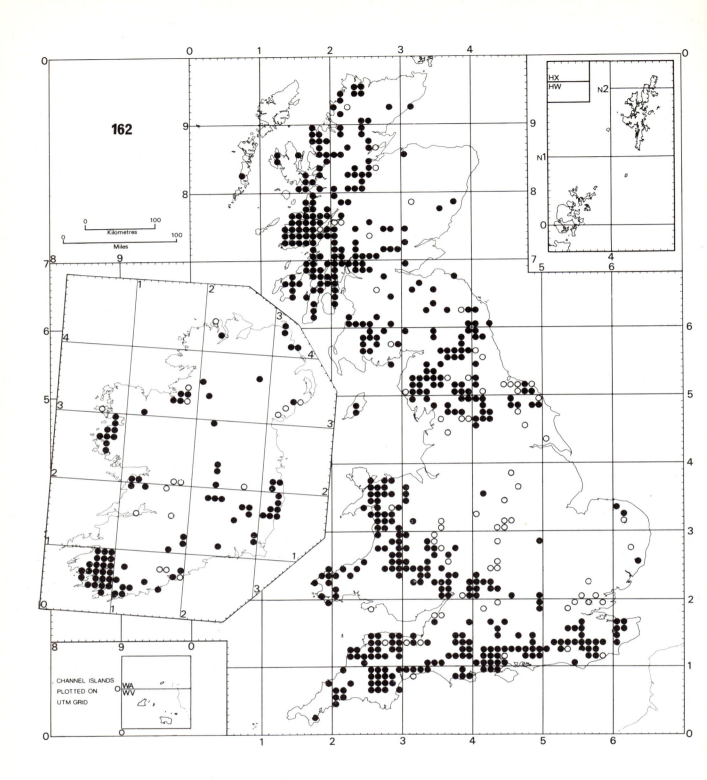

162 *Thelotrema lepadinum* (Ach.) Ach.

A species of smooth-barked boles and branches of broad-leaved trees, very rarely acid rocks. It grows on a wide variety of phorophytes, of which *Corylus*, *Fagus*, *Ilex* and *Quercus* are the most important. It is characteristic of a moderately shaded, more humid, often species-rich (in the west) variant of the *Graphidion scriptae*. In lowland Britain *T. lepadinum* is a valuable indicator species of old-established and undisturbed woodlands or parklands; in the highland zone it is widely distributed and also occurs on younger and smaller trees, usually in sheltered valleys. In Europe it is present from Greece to Scandinavia; it has also been recorded from many sub-tropical, as well as temperate, areas in the southern hemisphere.

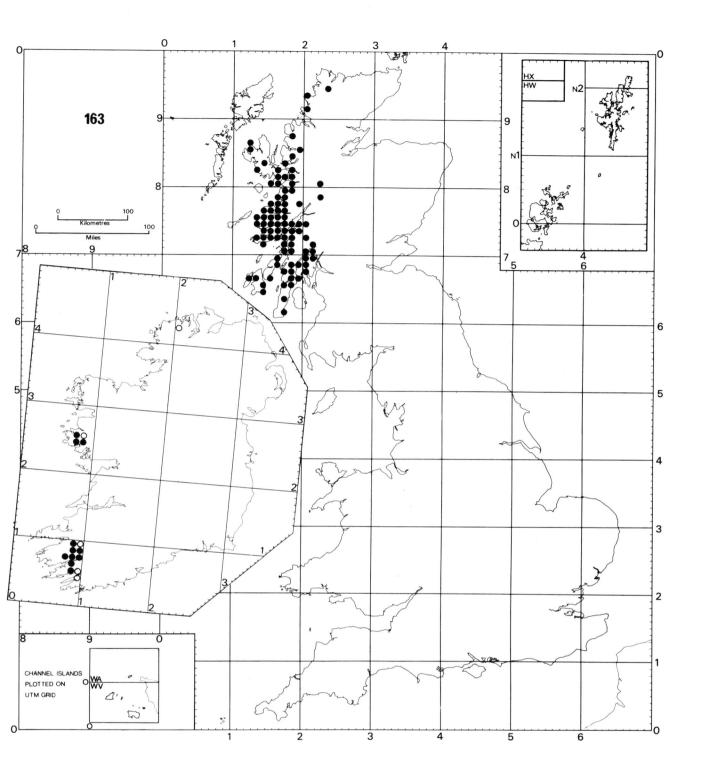

63 *Thelotrema subtile* Tuck.

A markedly euoceanic species on smooth-barked trees in lowland (below 60 m), sheltered, moist, well-wooded valleys and ravines or in lowland bogs and by sheltered streamsides; most frequent on *Corylus,* but also on *Fagus, Fraxinus, Ilex, Prunus padus, Quercus* and *Sorbus aucuparia.* It is an indicator species of an unusual species-rich facet of the *Graphidion,* sometimes frequent in restricted areas of west Scotland, south-west and west Ireland. Abroad only recorded from eastern USA and the Azores.

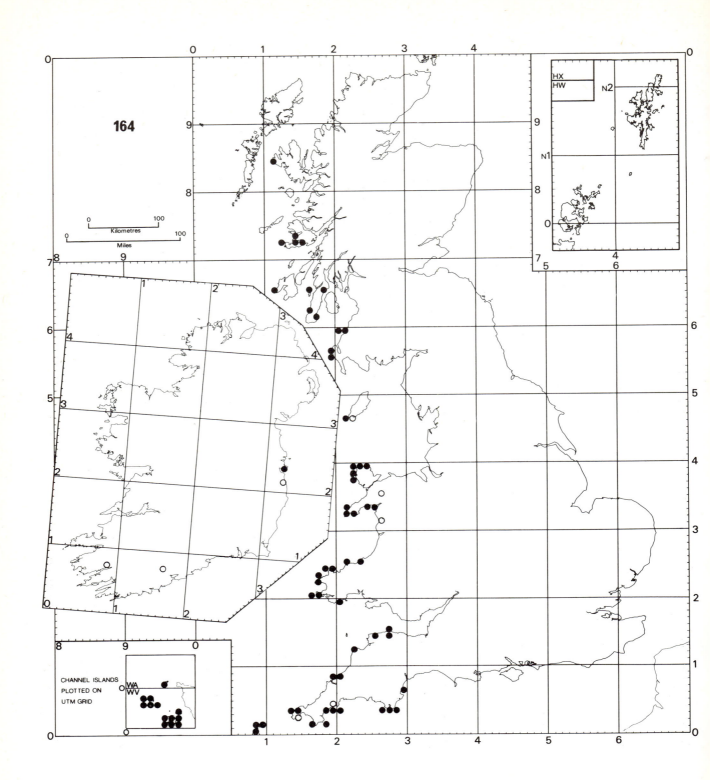

164 *Trapeliopsis wallrothii* (Flörke ex Sprengel) Hertel & G. Schneider

An exclusively xeric-supralittoral coastal species in western Britain, and characteristic of thin well-drained, more or less level, consolidated soils (pH 4.0-5.6) overlying acid rocks in sunny, exposed situations. *T. wallrothii* has been recorded as far north as Skye, but is most frequent in Wales (especially Anglesey) and south-west England. Also known from coastal and inland areas of France, Portugal, Spain, the Canary Islands and the Azores.

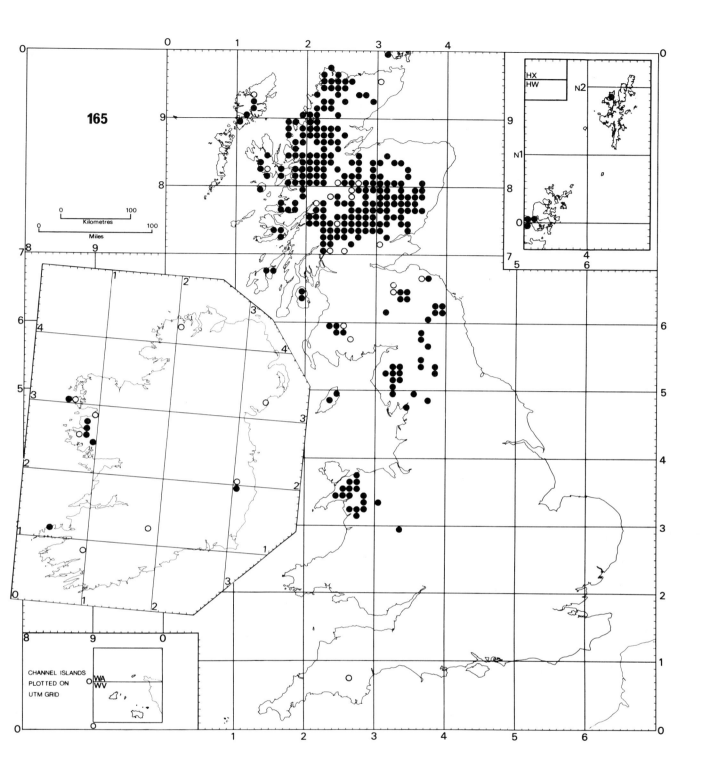

65 *Umbilicaria cylindrica* (L.) Delise ex Duby

A predominantly montane species associated with nutrient-poor, coarse-grained siliceous rocks in exposed, well-lit sites. The species is generally found between 200 and 1220 m, but descends to sea-level in north-west Scotland. It is a common species in the Scottish Highlands, becoming very local in west Scotland. In the Lake District, Cheviots and North Wales it is much more local and scarce, and it is now extinct on Dartmoor; in Ireland it is known from a few scattered localities, particularly in Connemara. *U. cylindrica* is an important species in the arctic-alpine *Rhizocarpetum alpicolae,* and dominant in the *Umbilicarietum cylindricae.* The species is circumboreal-montane, and is also known from the Himalayas and temperate regions of the southern hemisphere, including New Zealand and Tasmania.

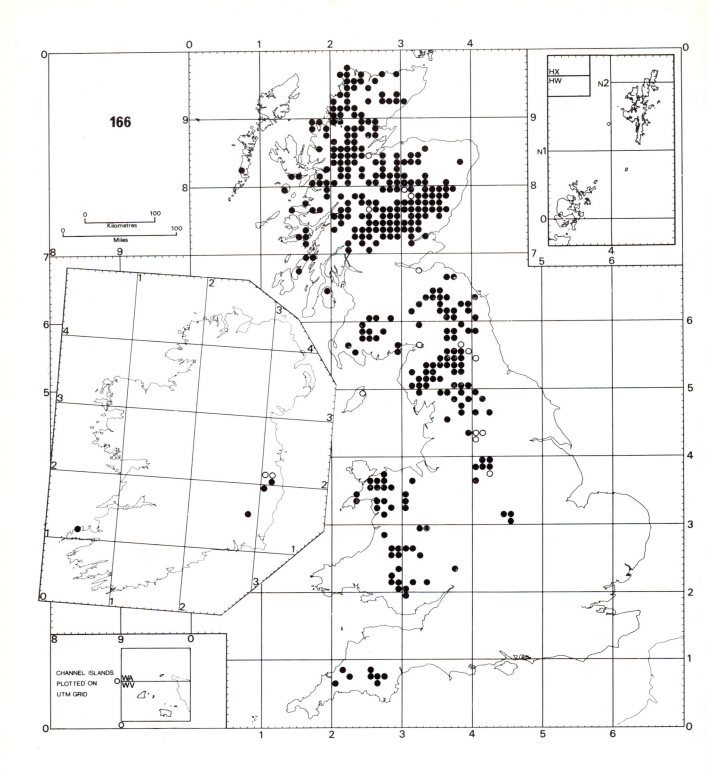

166 *Umbilicaria polyphylla* (L.) Baumg.

Predominantly a montane species, *U. polyphylla* is usually associated with moderately nutrient-enriched, silica-rich and coarse-grained rocks in exposed, well-lit sites. It is more tolerant of eutrophication than *U. cylindrica* (map 165), sometimes occurring on rocks manured by birds. It mainly grows between 200 and 1220 m, but in Scotland it descends to the sea coast, often on large boulders amongst *Calluna*. Its distribution resembles that of *U. cylindrica,* but it is more frequent than that species in northern England, north and south Wales and south-west England; there are outlying stations for the species in Charnwood Forest, Leicestershire; these sites, and those in the Peak District, show that the species is moderately pollution-tolerant. *U. polyphylla* is an important species in the associations *Rhizocarpetum alpicolae* and *Umbilicarietum cylindricae.* Abroad *U. polyphylla* is low arctic to circumboreal; it also occurs in southern South America, New Zealand and eastern Australia.

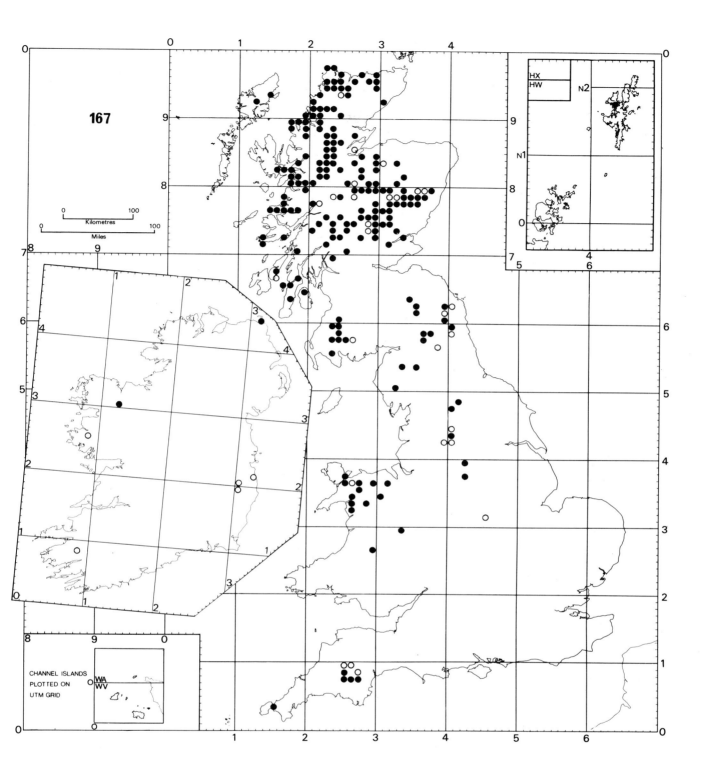

167 *Umbilicaria polyrrhiza* (L.) Fr.

U. polyrrhiza, like *U. polyphylla* (map 166) and *U. torrefacta* (map 168), is a widespread montane species on somewhat nutrient-enriched, silica-rich and coarse-grained rocks in exposed, well-lit sites. It chiefly grows on rock outcrops at altitudes between 200 and 1220 m but may also occur on large boulders and outcrops in moorland at sea-level. It is an important associate of the *Rhizocarpetum alpicolae* and the *Umbilicarietum cylindricae*. It is widespread but not as frequent as the other two species cited above; it is sparsely distributed in northern England and north Wales and is still locally abundant on some of the granite tors of Dartmoor. Reported from boreal and subarctic areas in both the north and south temperate areas; also on high mountains in central Europe and Asia.

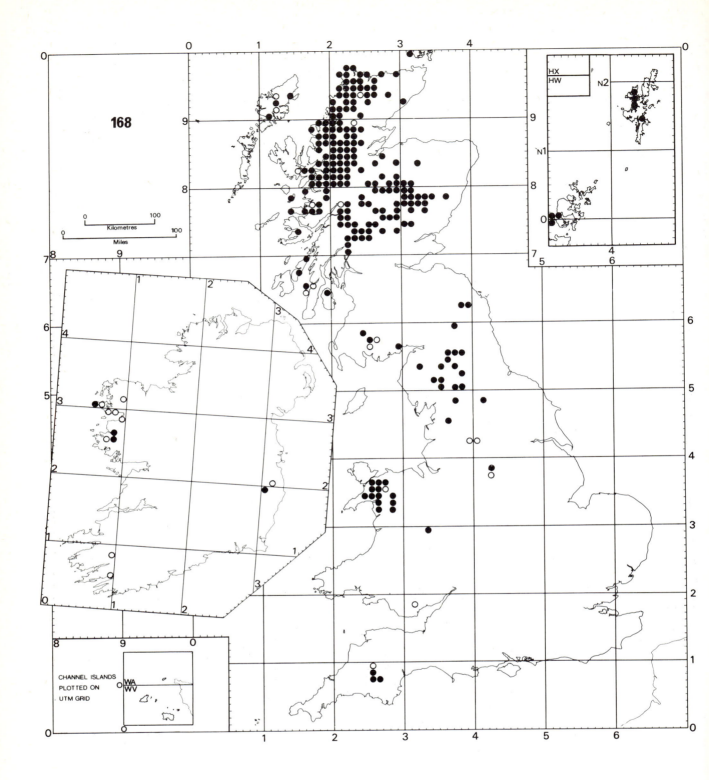

168 *Umbilicaria torrefacta* (Lightf.) Schrader

A widespread species of montane areas in Scotland, *U. torrefacta* usually colonises more or less nutrient-enriched, siliceous and coarse-grained rocks in exposed, well-lit sites. It is recorded from sea-level to 1230 m, and tends to favour more or less horizontal surfaces and damper sites than *U. cylindrica* (map 165) and *U. polyrrhiza* (map 167), with which it often grows. Like these species it occurs in the *Rhizocarpetum alpicolae* and *Umbilicarietum cylindricae*. It is local in northern England and locally common in north Wales; it is still present on the granite tors of Dartmoor. A circumboreal, subarctic species; also known on the higher mountains in southern Europe and North America.

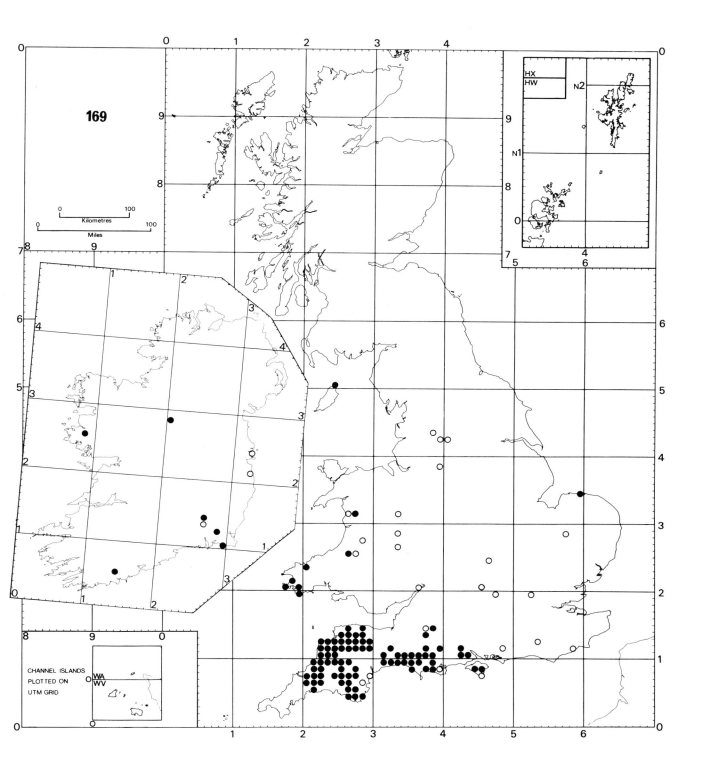

69 *Usnea articulata* (L.) Hoffm.

A predominant species of the *Usneetum articulato-floridae* var. *ceratinae* characteristic of the very well-lit xeric situations encountered on the uppermost branches or horizontal boughs of deciduous trees. It is now mainly confined to south-west England, and is still an exceptionally common and luxuriant species in the Bodmin and Holsworthy-Clovelly areas where it grows to over 1 m in length, colonizing hedges and even coniferous trees; also found straggling over mosses, other lichens and short vegetation on sand dunes (e.g. Isle of Man) or boulders. The species was first described from a 30 cm long specimen on *Corylus* collected in Burnley (Lancashire) in 1692 and was evidently formerly much more widespread in Britain; increasing air pollution had already eliminated it from much of lowland Britain by the end of the eighteenth century. One of the most sensitive species to sulphur dioxide air pollution and absent when mean winter levels exceed about 30 μg m^{-3}. Now extinct in many parts of Europe but still thriving in the clean air of Brittany; absent from Scandinavia. Also known from mountainous areas of east and north Africa, Saudi Arabia and the Atlantic and Mediterranean Islands. The British material uniformly contains fumarprotocetraric acid; extra-European specimens have an extremely diverse chemistry.

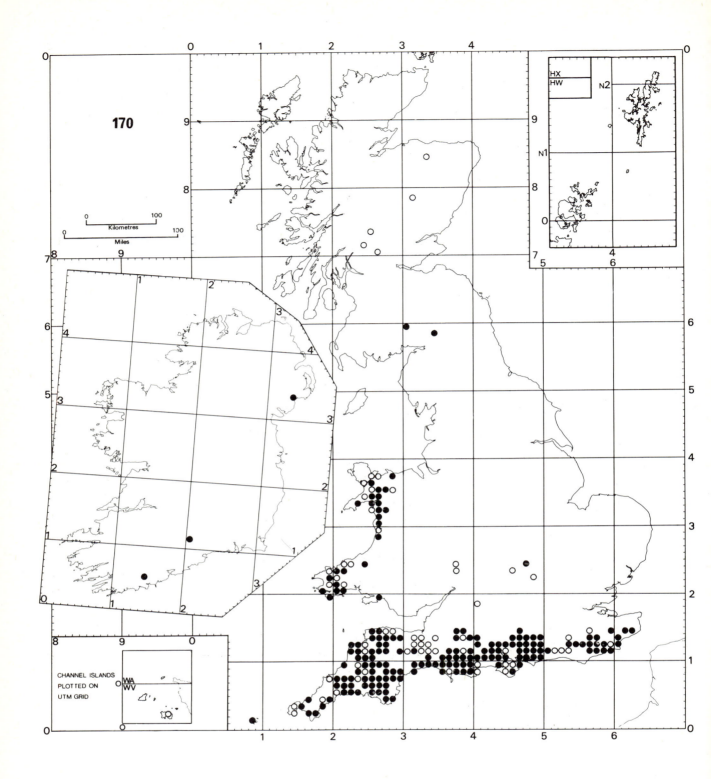

170 *Usnea ceratina* Ach.

An old woodland and parkland species on acid bark of old trees, including *Quercus* and *Fagus,* more rarely *Crataegus, Salix, Alnus* and *Castanea. U. ceratina* belongs to a spectacular community, the *Usneetum articulato-floridae* var. *ceratinae* characteristic of well-lit sites and optimally developed on the uppermost sloping or horizontal boughs in rather thin tree canopies; also on boles in more open humid situations such as at the edge of boggy clearings, lakesides and *Salix* carr; it is not, however, a species of wayside trees, except for the least polluted parts of Devon and Cornwall. Distinguished from other British *Usnea* species by the pendulous habit, dark grey-green colour, the coarse, white papillae, the pink and dense medullary tissue, and the presence of diffractaic acid. The species has a markedly southern and south-western distribution in Britain and is known also from north and south Wales. The most northerly record is in Dumfriesshire (Lochmaben, Lochwood), but there are a few scattered old records for the Central Highlands of Scotland. It is a very pollution sensitive (> 30 μg m^{-3}) species and was probably formerly much more widely distributed in the south Midlands, where a combination of air pollution, and destruction and drainage of suitable habitats has caused the species to die out. Its distribution in Ireland is uncertain, but it is probably present in most wooded areas in the south. Recorded from southern, western and central Europe, extending into extreme southern Scandinavia where it is very rare; reported, but in need of confirmation, from North America.

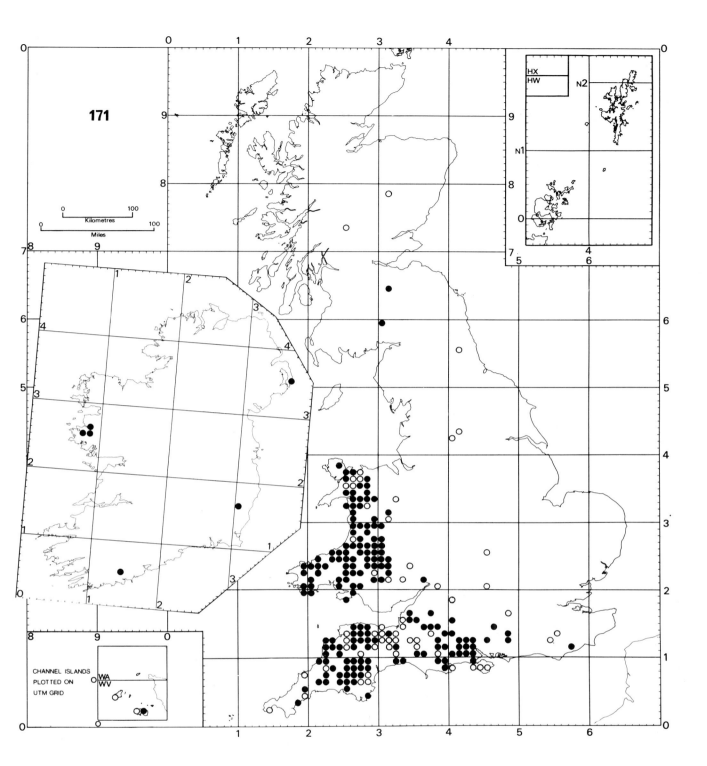

171 *Usnea florida* (L.) Wigg.

A distinctive species most frequent in the upper canopy of medium-aged to old trees in natural broad-leaved woods or plantations. The species is most frequent on the twigs and small branches of *Quercus,* but may occur on a wide range of phorophytes, including *Salix, Fraxinus, Castanea, Betula* and *Crataegus,* as well as shrubs, palings, orchards and fence posts. It belongs to the *Usneetum articulato-floridae* var. *ceratinae* in well-lit situations and the *Usneetum subfloridanae* where this occurs in the south of Britain; the latter association is a predominantly small-branch-twig association in very well-lit situations, preferring a slightly less acid bark than the former community. *U. florida* contains thamnolic acid and often alectorialic acid, the latter more or less concentrated in the apothecia. The species is the fertile counterpart to a similar chemotype of *U. subfloridana* which differs in having soralia with soredia and pseudoisidia. *U. florida* has a very similar distribution to *U. ceratina* (map 170), but is more frequent in Wales and rarer in Sussex and Kent. The most northern modern record is in Dumfriesshire (Lochmaben, Lochwood), but there are old verified records for the Central Highlands of Scotland. The species is European, extending to southern Scandinavia and Poland in the east; formerly in much of central Europe but, as in Britain, now much rarer due to the effects of air pollution and the destruction of suitable habitats.

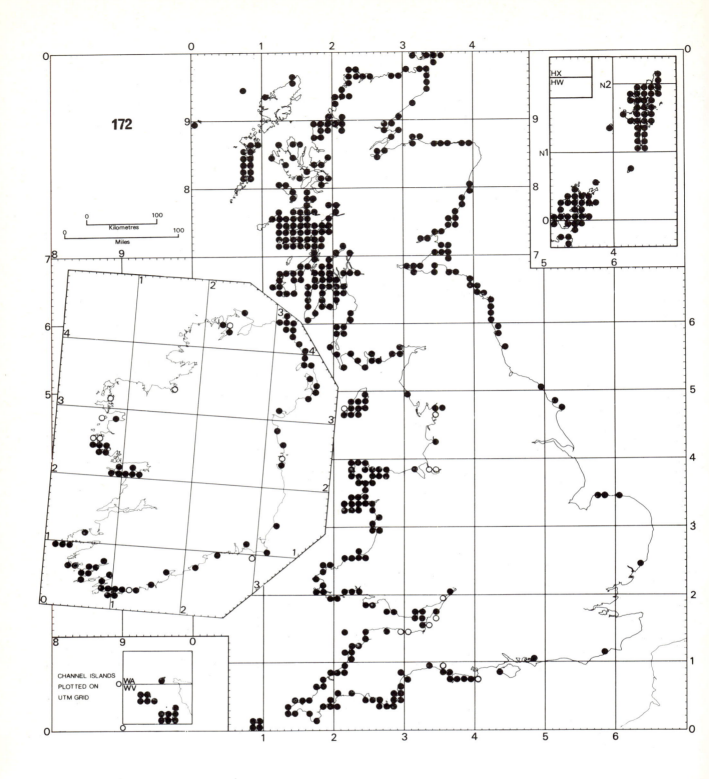

172 *Verrucaria maura* Wahlenb. ex Ach.

On siliceous and calcareous rocks, provided that they are not too soft or easily eroded, on shingle, cement and brickwork, animal shells and very rarely on wooden posts. A very common species which is widespread on seashores throughout the country, even extending into muddy areas when man-made hard substrata are available. Restricted to the littoral zone, its upper limit on the shore is used to define the boundary between terrestrial and marine biotic communities. It is abundant on sheltered shores, becoming less common on exposed rocks where it tends to be restricted to crevices. On the most exposed shores it penetrates 100 m or more up cliffs and can appear on buildings and rocks 1 km inland. It is tolerant of very low salinities and penetrates well into estuaries. The species is well developed on both sunny and shaded shores. It has a wide ecological tolerance and is morphologically highly variable. The species may have declined in abundance in silt-polluted situations such as Morecambe Bay, the Bristol Channel and north-east England. It has been well recorded and is conspicuous, but older literature records probably include other related species. It is widespread throughout polar and temperate regions, where its ecology is similar to that in Britain, but it becomes rare and confined to crevices in warmer climates. It is said to be a shade-loving plant in west Norway.

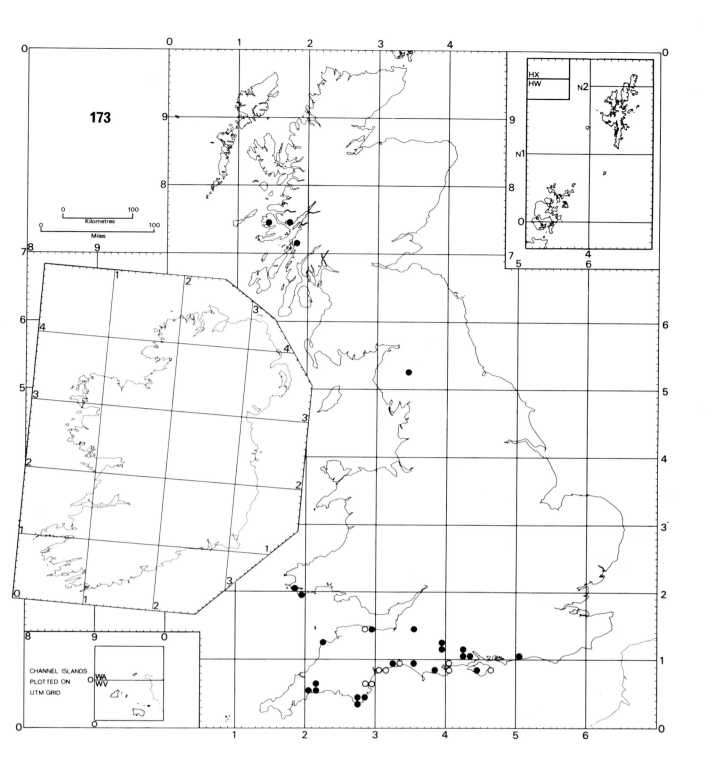

173 *Wadeana dendrographa* (Nyl.) Coppins & P. James

A species of old rough basic-barked trees, commonly *Fraxinus* and *Ulmus,* rarely *Quercus,* in sunny but moist clearings, in old woodlands or wayside sites, on rich alluvial or basic volcanic soils. The species probably belongs to a facies of the *Lobarion pulmonariae* which is rich in cyanophilous species such as *Collema* and *Leptogium;* it also occurs in the *Gyalectinetum carneoluteae.* It is a very local lichen, restricted to south and south-west England with outliers in Pembroke, Westmorland and west Scotland. Most localities are near the coast and the species appears to require an equable frost-free, warm habitat. It has become extinct in some stations due to reafforestation and pollution by inorganic fertilizers. Otherwise, only known from Portugal.

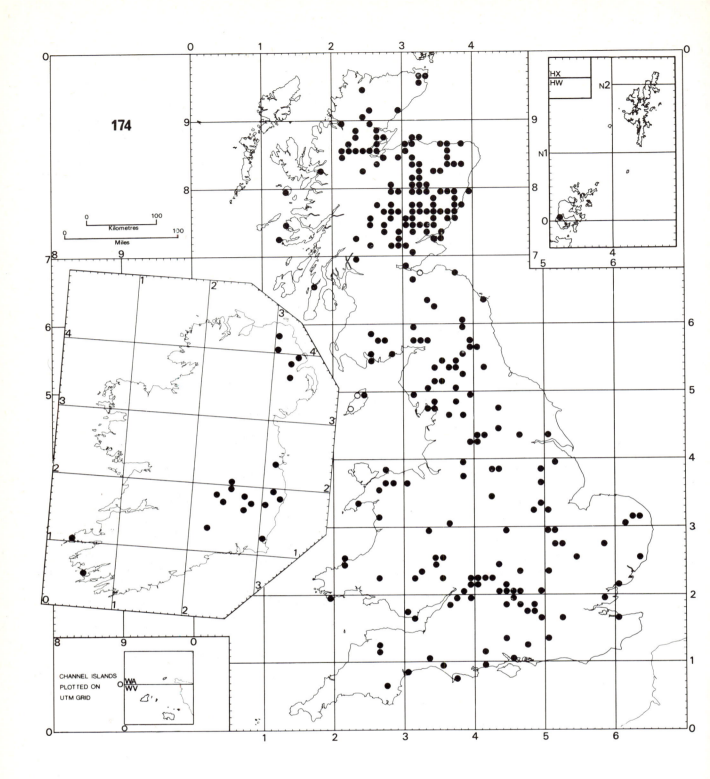

174 *Xanthoria elegans* (Link) Th.Fr.

Primarily a species of nutrient-enriched bird-perches on siliceous rocks in upland situations, occurring in the *Physcietum caesiae,* but also known on flint pebbles and wood. However, in the last fifteen years it seems to have markedly extended its range onto man-made substrates in lowland areas (cf. *Lecanora muralis,* map 67), including asbestos-cement, concrete, mortar, slate roofs, and walls; the amount of expansion is variable in different parts of the country, and an association with car parks and tourist haunts suggests that it can be dispersed by man. Evidently fairly tolerant of sulphur dioxide pollution, so further expansion on man-made substrates can be anticipated. Almost cosmopolitan, broadly circumpolar in both hemispheres, ranging from extreme arctic localities through boreal and temperate zones to the Antarctic; reported from Africa, but apparently absent from Australia.

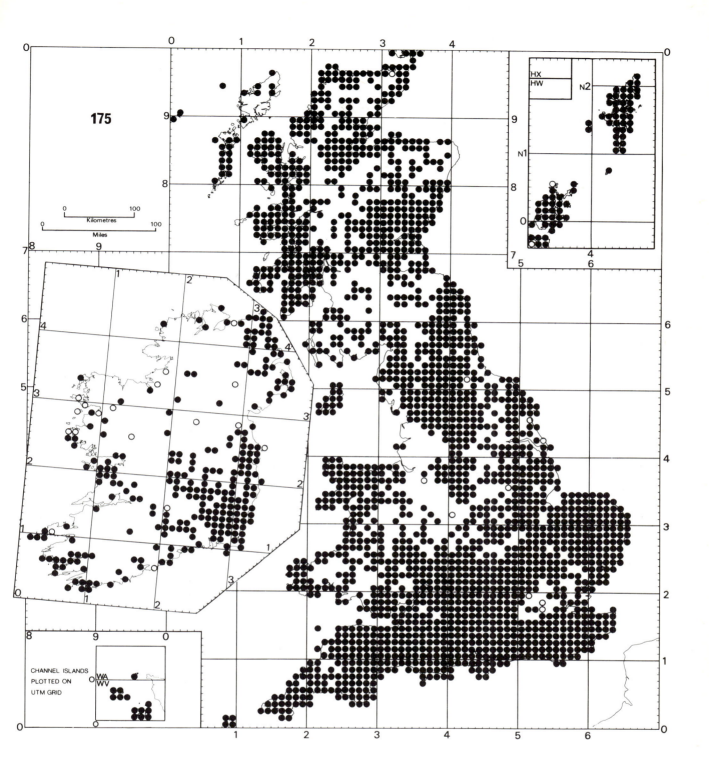

75 *Xanthoria parietina* (L.) Th.Fr.

One of the commonest British macrolichens, it occurs in a number of different communities in a wide range of natural and man-made habitats with a pH of over 6. Characteristic of the *Xanthorion parietinae* which requires nutrient-rich, enriched, or hypertrophicated substrates. Recorded on bark, siliceous or calcareous rocks, asbestos-cement, brick, bone, leather, bitumen roofs, glass, iron, flint, etc. in the British Isles. Appearing greyish in shaded situations, and vivid orange with somewhat narrower lobes when on rocks in the submesic to xeric supralittoral zones on rocky shores. Very tolerant of air pollution, especially on asbestos-cement where it withstands mean annual sulphur dioxide levels to about 125 μg m^{-3}, and tolerating up to about 80 μg m^{-3} on nutrient-rich bark. Reputed to be cosmopolitan, but not as tolerant of extreme arctic or antarctic situations as *X. elegans*. It extended its range considerably in Finland after the late nineteenth century, and similar extensions may have occurred in the British Isles and elsewhere in Europe, but are not yet documented. Also widespread in the Mediterranean region, north and east Africa, the Middle East and the temperate southern hemisphere where nutrient-rich bark communities predominate on trees and shrubs.

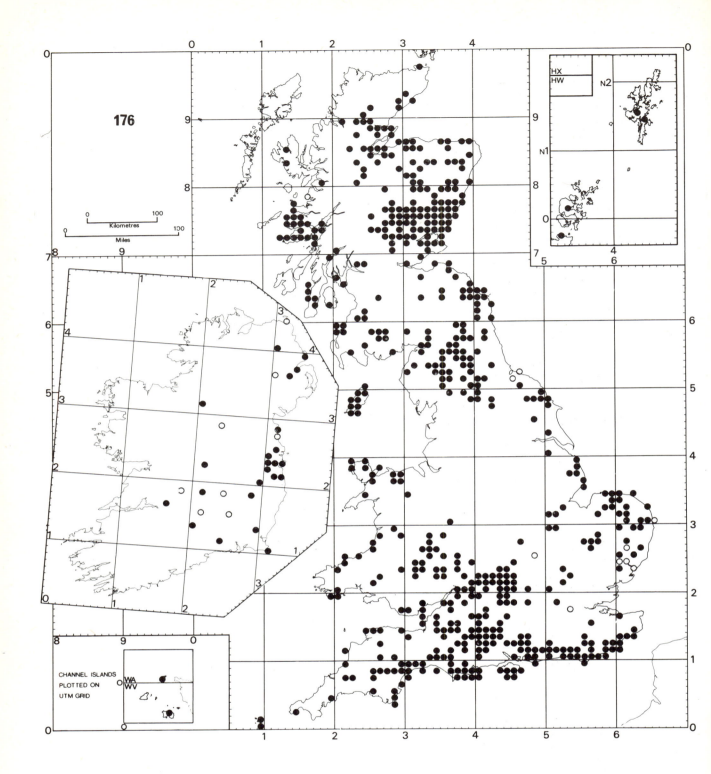

176 *Xanthoria polycarpa* (Hoffm.) Rieber

A fairly common species, mainly on twigs of well-lit trees and bushes, especially *Fraxinus* and *Ulmus,* particularly in well-ventilated, usually nutrient-rich or enriched situations. Less commonly found on trunks of trees with basic bark in the *Physcietum ascendentis.* On twigs it occurs in various communities, excluding the *Teloschistetum flavicantis,* often associated with *Physcia adscendens, P. aipolia, P. semipinnata, Parmelia exasperata, Rinodina sophodes* and various green algae. Not occurring where mean winter sulphur dioxide levels exceed about 60 μg m⁻³, but recently discovered on dead elm twigs in north-west London. Also sensitive to hypertrophication by modern fertilizer drift. Particularly common near the sea in southern Britain, but also quite frequent inland; rare or absent in the very exposed hyperoceanic areas of north-west Scotland. Widespread and common throughout Europe from Norway southwards. A southern boreal and temperate to subtropical species, circumpolar in the northern hemisphere but with some continental tendencies. Distribution in the southern hemisphere uncertain due to confusion with other species.

INDEX

APPENDIX A

Checklist of published distribution maps of British lichens

REFERENCES

1. **Bailey, R.H.** 1974. Distribution maps of lichens in Britain. Map 8. *Lichenologist,* **6,** 169-171.
2. **Bailey, R.H. & James, P.W.** 1977. Distribution maps of lichens in Britain. Map 23. *Lichenologist,* **9,** 175-179.
3. **Brightman, F.H., James, P.W. & Rose, F.** 1973. Distribution maps of lichens in Britain. Map 6. *Lichenologist,* **5,** 476-477.
4. **Coppins, B.J.** 1976. Distribution patterns shown by epiphytic lichens in the British Isles. In: *Lichenology: progress and problems,* edited by D.H. Brown, D.L. Hawksworth and R.H. Bailey, 249-278. London: Academic Press.
5. **Coppins, B.J. & James, P.W.** 1974. Distribution maps of lichens in Britain. Maps 9 and 10. *Lichenologist,* **6,** 172-177.
6. **Coppins, B.J. & James, P.W.** 1978. New or interesting British lichens II. *Lichenologist,* **10,** 179-207.
7. **Coppins, B.J. & James, P.W.** 1979. New or interesting British lichens IV. *Lichenologist,* **11,** 139-179.
8. **Dobson, F.S. & Hawksworth, D.L.** 1976. *Parmelia pastillifera* (Harm.) Schub. & Klem. and *P. tiliacea* (Hoffm.) Ach. in the British Isles. *Lichenologist,* **8,** 47-59.
9. **Earland-Bennett, P.M.** 1975. *Lecanora subaurea* Zahlbr., new to the British Isles. *Lichenologist,* **7,** 162-167.
10. **Gilbert, O.L.** 1975. Distribution maps of lichens in Britain. Maps 19 to 22. *Lichenologist,* **7,** 181-192.
11. **Gilbert, O.L.** 1978. *Fulgensia* in the British Isles. *Lichenologist,* **10,** 33-45.
12. **Hawksworth, D.L.** 1972. Regional studies in *Alectoria* (Lichenes) II. The British species. *Lichenologist,* **5,** 181-261.
13. **Hawksworth, D.L.** 1972. The natural history of Slapton Ley Nature Reserve. IV. Lichens. *Fld Stud.,* **3,** 535-578.
14. **Hawksworth, D.L.** 1973. Some advances in the study of lichens since the time of E.M. Holmes. *Bot. J. Linn. Soc.,* **67,** 3-31.
15. **Hawksworth, D.L.** 1973. Ecological factors and species delimitation in the lichens. In: *Taxonomy and ecology,* edited by V.H. Heywood, 31-69. London: Academic Press.
16. **Hawksworth, D.L.** 1974. Lichens and indicators of environmental change. *Environ. & Change,* **2,** 381-386.
17. **Hawksworth, D.L.** 1974. Man's impact on the British fauna and flora. *Outl. Agric.,* **8,** 23-28.
18. **Hawksworth, D.L.** 1980. Lichens of the south Devon coastal schists. *Fld Stud.,* **5,** 195-227.
19. **Hawksworth, D.L. & Chapman, D.S.** 1971. *Pseudevernia furfuracea* (L.) Zopf and its chemical races in the British Isles. *Lichenologist,* **5,** 51-58.
20. **Hawksworth, D.L., Coppins, B.J. & Rose, F.** 1974. Changes in the British lichen flora. In: *The changing flora and fauna of Britain,* edited by D.L. Hawksworth, 47-78. London: Academic Press.
21. **Hawksworth, D.L. & James, P.W.** 1974. Distribution maps of lichens in Britain. Map 16, *Lichenologist,* **6,** 194-196.
22. **Hawksworth, D.L. & Rose, F.** 1976. *Lichens as pollution monitors.* London: Edward Arnold.
23. **Hawksworth, D.L., Rose, F. & Coppins, B.J.** 1973. Changes in the lichen flora of England and Wales attributable to pollution of the air by sulphur dioxide. In: *Air pollution and lichens,* edited by B.W. Ferry, M.S. Baddeley and D.L. Hawksworth, 330-367. London: Athlone Press.
24. **James, P.W.** 1974. Distribution maps of lichens in Britain. Map 17. *Lichenologist,* **6,** 197-199.
25. **James, P.W.** 1975. The genus *Gyalideopsis* Vezda in Britain. *Lichenologist,* **7,** 155-161.
26. **James, P.W.** 1977. Distribution maps of lichens in Britain. Maps 24 and 25. *Lichenologist,* **9,** 181-187.
27. **James, P.W. & Hawksworth, D.L.** 1974. Distribution maps of lichens in Britain. Map 11. *Lichenologist,* **6,** 178-180.
28. **James, P.W. & Rose, F.** 1973. Distribution maps of lichens in Britain. Maps 2 to 5. *Lichenologist,* **5,** 467-475.
29. **James, P.W. & Rose, F.** 1973. Distribution maps of lichens in Britain. Map 7. *Lichenologist,* **5,** 478-480.
30. **James, P.W. & Rose, F.** 1974. Distribution maps of lichens in Britain. Maps 12 to 15. *Lichenologist,* **6,** 181-193.

31. **Seaward, M.R.D.** 1975. Contributions to the lichen flora of south-east Ireland. *Proc. R. Ir. Acad., B,* **75**, 185-205.

32. **Seaward, M.R.D.** 1976. Performance of *Lecanora muralis* in an urban environment. In: *Lichenology: progress and problems,* edited by D.H. Brown, D.L. Hawksworth and R.H. Bailey, 323-357. London: Academic Press.

33. **Sheard, J.W.** 1967. A revision of the lichen genus *Rinodina* (Ach.) Gray in the British Isles. *Lichenologist,* **3**, 328-367.

APPENDIX B

LIST OF FIELDWORKERS

Allen, N.V.
Alvin, K.
Arnold, M.A.
Ashby, R.T.
Bailey, R.H.
Barclay, D.J.
Bates, J.W.
Bennell, A.P.
Bennell, F.M.
Birks, H.J.B.
Blanchard, D.L.
Bowen, H.J.M.
Brand, A.M.
Brightman, F.H.
Brinklow, J.E.
Brinklow, R.K.
Broad, K.
Brown, D.H.
Burnet, A.M.
Campbell, I.S.C.
Chandler, J.H.
Chapman, D.S.
Cleden, J.L.
Conolly, A.P.
Coppins, B.J.
Corner, R.W.M.
Cox, R.
Crawley, M.J.
Crittenden, P.D.
Davey, S.R.
Davies, F.B.M.
Dixon, N.E.
Dobson, F.S.
Dudley-Smith, R.
Duncan, U.K.
Earland-Bennett, P.M.
Evans, I.M.
Farrar, J.F.
Fenton, A.F.G.
Ferry, B.W.
Fildes, J.
Finney, I.D.
Fletcher, A.
Fox, B.W.
Garrett, R.R.M.
Gilbert, O.L.
Gomm, F.R.
Graham, G.G.
Guiterman, J.D.
Hackney, P.
Hambler, D.J.
Harrold, P.
Hawksworth, D.L.
Hazelwood, E.
Heath, W.
Henderson, A.
Hickmott, M.
Hill, D.J.
Hinton, V.A.
Hiscock, K.
Hitch, C.J.B.
Holligan, P.M.
Hunt, D.J.

Huxley, C.
James, P.W.
Johnson, M.G.
Jones, I.
Jones, W.E.
Joyce, A.R.
Kay, Q.O.N.
Kerr, A.J.
Klein, J.
Lambley, P.W.
Laundon, J.R.
Lindsay, D.C.
Manning, S.A.
McCarthy, P.M.
Mellor, D.
Metcalfe, R.J.A.
Millar, R.O.
Mitchell, M.E.
Moon, H.P.
Moore, C.C.
Morgan-Huws, D.
Morton, O.
Mothersill, C.
Myall, J.
Neff, M.J.
Nelson, B.H.
Nelson, E.C.
Nethercott, P.J.M.
Neville-Smith, V.
Noon, R.A.
O'Hare, G.P.
Pentecost, A.R.
Peterken, J.G.
Pope, C.R.
Prince, C.R.
Proctor, M.C.F.
Ranwell, D.S.
Redshaw, E.J.
Richardson, C.B.
Richardson, D.H.S.
Roper-Lindsay, J.
Rose, C.I.
Rose, F.
Ross, K.W.
Ruddock, K.
Sayer, P.
Scannell, M.J.P.
Seaward, M.R.D.
Shaw, P.J.
Sheard, J.W.
Showell, J.P.
Shrimpton, A.
Shuttleworth, S.
Sipman, H.
Skinner, J.F.
Sowter, F.A.
Spooner, M.
Stewart, P.R.
Stonehouse, B.
Swinscow, T.D.V.
Tallowin, S.N.
Telford, M.
Titterington, R.

Tittley, I.
Topham, P.B.
Turpitt, L.
Wade, A.E.
Walker, F.J.
Walkinshaw, D.A.
Wallace, E.C.
Wallace, N.
Watling, R.
Watson, K.
Webster, M.M.
Wilkie, I.C.
Woods, R.G.
Wyatt, T.